PROJECT AIR FORCE

T0146373

Articulating the Effects of Infrastructure Resourcing on Air Force Missions

Competing Approaches to Inform the Planning, Programming, Budgeting, and Execution System

Patrick Mills, Muharrem Mane, Kenneth Kuhn, Anu Narayanan,
James D. Powers, Peter Buryk, Jeremy M. Eckhause, John G. Drew,
Kristin F. Lynch

Prepared for the United States Air Force

For more information on this publication, visit www.rand.org/t/RR1578

Library of Congress Cataloging-in-Publication Data is available for this publication.

ISBN: 978-0-8330-9677-7

Published by the RAND Corporation, Santa Monica, Calif.

© Copyright 2017 RAND Corporation

RAND® is a registered trademark.

Support RAND

Make a tax-deductible charitable contribution at

www.rand.org/giving/contribute

www.rand.org

Preface

The success of any U.S. Air Force mission depends on the availability and performance of its supporting infrastructure. But because of the complexities in the way that myriad infrastructure assets support a wide range of missions, the proper level of funding for infrastructure maintenance can be difficult to establish or defend, and the detrimental effects of chronic underfunding on mission capability and readiness may not become apparent for several years.

This report discusses several analytic approaches for linking infrastructure resources to readiness and for articulating the effect of infrastructure underfunding in the Air Force Program Objective Memorandum (POM) process. This analysis was performed in support of a RAND Project AIR FORCE (PAF) project titled "Infrastructure Resources to Readiness" sponsored by Maj Gen Theresa Carter, former Deputy Chief of Staff, Logistics, Engineering, and Force Protection, and was conducted within the Resource Management Program of PAF.

This report should interest personnel in the areas of infrastructure, logistics, and programming and budgeting in the Air Force and in the broader defense community.

RAND Project AIR FORCE

RAND Project AIR FORCE (PAF), a division of the RAND Corporation, is the U.S. Air Force's federally funded research and development center for studies and analyses. PAF provides the Air Force with independent analyses of policy alternatives affecting the development, employment, combat readiness, and support of current and future air, space, and cyber forces. Research is conducted in four programs: Force Modernization and Employment; Manpower, Personnel, and Training; Resource Management; and Strategy and Doctrine. The research reported here was prepared under contract FA7014-06-C-0001.

Additional information about PAF is available on our website: http://www.rand.org/paf/.

This report documents work originally shared with the U.S. Air Force on July 15, 2014. The draft report, issued on October 22, 2015, was reviewed by formal peer reviewers and U.S. Air Force subject-matter experts.

Contents

Summary

The Air Force civil engineering community has found that its methods for articulating infrastructure funding needs and mission impacts in the Program Objective Memorandum (POM) process are insufficient, and it is in the process of investigating alternatives. In fiscal year (FY) 2014, the Air Force Civil Engineer asked RAND Project AIR FORCE (PAF) to investigate how the Air Force might articulate the effects of sustainment, restoration, and modernization (SRM)[1] underfunding on readiness to ensure adequate funding to support these activities.

This analysis explores the relationship between Air Force infrastructure management and mission readiness and capability. The goal is to identify methodological approaches and data requirements for articulating and quantifying these links and enabling the Air Force to answer the question: What is the effect of funding infrastructure below stated requirements?

Background

The success of any U.S. Air Force mission depends on the availability and performance of its supporting infrastructure. In some cases, the linkage between infrastructure and mission capability is clear. For example, a closed runway directly affects sortie generation capability. Most of the time, however, the connection is far less direct. While few would dispute that a poorly maintained runway increases aircraft wear and tear, eventually yielding greater fleet repair costs and reduced availability, such effects can be difficult to quantify or trace back to the underlying causes. As a result, the proper level of funding for infrastructure maintenance can be difficult to establish or defend, and the detrimental effects of chronic underfunding on mission capability and readiness may not become apparent for several years.[2]

Infrastructure degrades over time and with use, and infrastructure maintenance and repair activities keep it in good working order for its intended service life. The Air Force Civil Engineer and base civil engineers (BCEs) sustain the array of infrastructure assets and systems using a range of small- and large-scale activities, most of which fall under the umbrella of SRM. SRM includes activities ranging from preventive maintenance tasks, to periodic activities like regular roof replacement, to repairing damage of many kinds, to upgrading components or whole facilities to conform to recent standards.[3]

---

[1] In commercial practice, this is usually referred to as maintenance, renovation, and reconstruction (MR&R).

[2] In this report, we use the terms *readiness*, *mission capability*, and *mission performance* more or less interchangeably. *Readiness* often has very specific meanings, such as the financial accounts that underwrite training activities, or the actual readiness reporting systems and output metrics (e.g., C-ratings). Here, we use all these terms in a fairly generic sense of the ability to perform a given set of tasks or objectives.

[3] For more expansive definitions, see Air Force Instruction 32-1032, 2014.

Current Methods Used to Estimate SRM Funding Requests

There is no "right" way to define enough funding for infrastructure maintenance and repair, and there is no "right" way to quantify it for the purposes of budgeting. There exist a number of competing methods to estimate this need for annual funding, from very coarse to very detailed.

The Air Force uses a combination of methods, including bottom-up gathering of base-level needs for day-to-day maintenance and larger projects, as well as high-level parametric cost models (which are also commonly used in U.S. Department of Defense [DoD] and commercial practice). These provide very high-level estimates of needs and do not articulate the mission performance gained from a funding level (beyond "keep the infrastructure in good working order") nor mission risk incurred by accepting a lower level of funding.

Infrastructure degradation in DoD and the United States more broadly is deep and widespread. This is evidenced by visible deterioration, component failures, and fragility in the face of natural disasters, and explained by chronic underfunding of infrastructure maintenance budgets. The Air Force has also had chronic underfunding of infrastructure maintenance budgets, though tangible examples of deterioration and failure are usually anecdotal.

Alternative Approaches

To answer the question "What is the effect of funding infrastructure below stated requirements?" we assessed three alternative approaches: a project scorecard approach, an approach based on mission outcome metrics, and an approach based on composite risk metrics. We identified these competing analytic approaches for analyzing and presenting mission risk (whether specifically for linking infrastructure to mission or not) by reviewing relevant literature from academia, commercial practices and case studies, and DoD policies and practices. We also discussed a range of approaches with Air Force and other service personnel involved in engineering and/or POM deliberations. Then we assessed the strengths, weaknesses, and relative implementation burden of each approach. We also explored ways to mitigate the weaknesses of each approach to make them most useful in the Air Force context. Finally, we identified steps the Air Force can take to implement these concepts and to improve its ability to develop a systematic, evidence-based case for SRM funding within the POM process more generally.

Table S.1 summarizes the three approaches we identified in terms of the nature of their main output, the steps to produce them, and their strengths and weaknesses.

Table S.1. Summary of Three Alternative Approaches

Attribute	Project Scorecard	Mission Outcome Metrics	Composite Risk Metrics
Output	• List of projects up for deliberation with affected missions	• Quantification of future mission performance based on funding levels	• Expected composite risk ratings based on levels of funding
Steps	• Develop and prioritize project list from base-level inputs • Develop project tradespace (groups of projects) based on levels of funding • Apply mission areas to projects in tradespace • Mission owners advocate for projects in their purview	• Develop mission metrics for a single mission • Link infrastructure assets to mission outputs with logic model • Develop mathematical model quantifying relationships between key assets and mission outputs • Impact of projects under consideration quantified in terms of mission metrics	• Assess mission performance (i.e., condition, function) of infrastructure assets • Develop and apply mission performance thresholds for infrastructure assets • Translate mission performance to risk metrics • Develop composite risk profiles based on levels of funding
Strengths	• Concrete, focus on single projects • Puts onus on mission owners • Little additional investment	• Concrete • Mission-specific • Output-oriented • Can identify unique input/output relationships • Compatible with displaying long-term implications of underfunding	• Risk framework already exists • Data systems support this approach • Compatible with displaying long-term implications of underfunding
Weaknesses	• Could challenge bandwidth of decisionmakers • Could devolve to project rather than mission focus • Could default to decisionmaking biases • Near-term perspective limits ability to express long-term implications of underfunding	• Potentially costly to implement • Missions require separate models • Still requires cross-mission assessment/trades • Not all missions may be amenable • Excludes more distant infrastructure activities like "municipality," personnel support	• Metrics are more abstract • Requires investment of time and manpower to populate data

Project Scorecard

A significant portion of overall infrastructure spending goes to what are called "projects": activities that are beyond the scope of regular sustainment activities, usually because an asset is at a critical point—that is, it has been neglected for a long time, there is an impending failure, or the repair needs of the infrastructure asset exceed the base's organic repair capabilities (e.g., replacing the roof on a building). A project could be a single activity for a single asset, a group of activities for a single asset, or multiple activities involving multiple assets. Funding for small projects is controlled at the base level; funding for large projects is allocated through a formal, enterprise-level prioritization model, which results in a single integrated priority list (IPL).

We found that some infrastructure spending is essentially fixed and some is variable, with respect to the overall level of funding. The variable component mainly comprises projects, both large and small. Thus, when a total level of infrastructure funding is being debated, what is really in play is a list of projects of varying criticality (as indicated by the IPL), or, more specifically, the subset of those projects within the funding levels being considered, which we call the *project tradespace*.

Take, for example, a $2 billion funding request. If fixed costs for infrastructure funding (manpower, utilities, etc.) equal $1 billion, the request asks for another $1 billion that would ultimately be allocated to projects. If the ensuing debate considered whether to grant 80 or 90 percent of the total $2 billion request, the amount in play for projects on the IPL, i.e., the *project tradespace*, is only $0.6–0.8 billion, or 60–80 percent of the total request for projects. The net result of the debate is that projects within the funding threshold are funded, and those outside the threshold are delayed.

While the IPL is not presented in the current POM process, the project scorecard approach we assessed entails presenting a portion of the IPL, having decisionmakers review the list of projects within the project tradespace, and having mission owners explain the mission impact of deferring them.

A scorecard—a common tool in multi-criteria decisionmaking—is simply a table of options (usually in rows) with various criteria (usually in columns), and some scoring of each option along each criteria. Projects could be presented with or without additional information about the infrastructure asset's condition or mission criticality; the operative component is having mission owners advocate for them.

One strength of this approach is that a project (even when it encompasses multiple actions) is usually concrete and easy to understand or envision. Also, mission owners arguably know best the mission risk of deferring action. Third, this approach is simple, in that it requires little additional data gathering or processing.

There are several potential hazards involved in presenting individual projects. First, there could be an overwhelming amount of information and complexity in the project tradespace. There could be only a few projects up for debate, or dozens or hundreds, each with different kinds of mission impact.

Second, the mission impact of some projects may simply not be that compelling. Certainly, if the commander of Air Combat Command (ACC) argues that reduced funding would delay the repair of a key runway for a fighter pilot training base, or a maintenance hangar in need of renovation, the potential impacts are obvious, if sometimes uncertain. But our analysis suggests that many assets (excepting the most critical) could have mission risk or impact that is simply hard to intuit or internalize.

If the type and amount of information presented in the POM deliberations is, in fact, overwhelming, the deliberations could devolve into picking apart individual projects, scrutinizing the prioritization model itself, or trying to make smaller trades among projects that seem intuitively appealing.

Finally, the prioritized project list is only developed for near-term projects, which limits one's ability to express the long-term implications of underfunding.

Mission Outcome Metrics

The mission outcome metrics approach entails choosing useful mission metrics, then building logic and mathematical models to link and quantify the effects of infrastructure funding on these mission outcomes. Chapter Four contains a case study example of this approach that we conducted for Columbus Air Force Base (AFB), which does undergraduate pilot training on several aircraft types.

Like the project scorecard, well-designed mission outcome metrics are concrete and relatable: sorties generated, pilots graduated, etc. They are also mission-specific, so they are very tangible. They are also output-oriented, targeting what operators and decisionmakers are interested in and find compelling. More sophisticated models may also reveal interesting relationships, such as a knee in the curve where mission performance drops precipitously. In addition, models like these can capture interactions among a number of variables that may be too difficult to intuit without such an aid. And this approach can be used to express long-term impacts of underfunding.

The mission outcome metrics approach also has several weaknesses. First, these mission outcome models can be costly, in both time and manpower, to develop. Second, each mission (e.g., flying aircraft, space, cyber) potentially requires a separate model. Third, not all missions may be amenable to this kind of analysis. Fourth, this approach probably excludes more infrastructure assets and activities that may be more distant from individual missions, such as functions of the base as a "municipality." Finally, this approach still may require integration across projects and bases that all contribute to the same mission, which could require additional modeling.

Composite Risk Metrics

The composite risk metrics approach entails gathering and synthesizing data about infrastructure performance (using metrics like condition and functionality), applying performance thresholds based on user needs, and translating those ratings to some kind of holistic risk framework.

One advantage of this approach is that the Air Force's data systems (Sustainment Management Systems [SMSs]) already have some, if not all, of the data-handling capability needed. They are not all completely populated, but some of the Air Force's current information systems are designed to serve functions like this. To the degree that these data can be exported and synthesized, they can be leveraged to tie to a risk framework. AF/A9 has developed a risk framework that has already been applied in the Air Force, and so could be used for that purpose.

Like the mission metrics approach, the composite risk metrics approach can be used to express long-term impacts of underfunding.

While the AF/A9 risk framework provides a ready-made language to communicate to the AFCS, one weakness is that this is the most abstract of the three approaches we explored. One cannot see and touch stoplight metrics or risk indexes. It is often hard to differentiate moderate differences in risk in frameworks like this, and two options that appear to fall into the same category may have important differences.

Another weakness is that, while the Air Force's data systems provide the structure and machinery to process and output much of the needed data, those data must first be gathered. The systems are not yet fully populated, and incomplete data (i.e., not all bases, not all facilities on each base) could prevent valid analysis from being produced. The Air Force faces a question of cost-benefit trade-offs as to how much additional time and effort to put into populating these systems.

In this report, we also explore ways to mitigate the weaknesses of each approach and potentially combine some of the more attractive features. Our assessment of these three approaches led us to several conclusions and recommendations.

Conclusions

There are several viable approaches the Air Force can take to articulate mission impact; each has very different strengths, weaknesses, and implementation burden. All three approaches we reviewed are widely used in public and corporate decisionmaking, as well as policy analysis. We believe all three approaches may have a place in the Air Force as it transitions away from the status quo, though choosing a path ahead will require more thought and collaboration with infrastructure users and AFCS decisionmakers, and implementing that approach will likely require gathering more information.

That said, **the infrastructure-to-mission mapping exercise appears to have several potential side benefits.** These maps can reveal and clarify critical linkages. It could be useful to incorporate these products in a base's development of its contingency response plan (CRP) requirements (Air Force Instruction 10-211, 1998), or leverage them to inform currently implemented metrics, such as the Mission Dependency Index (MDI).

Solid risk analysis and communication are necessary, but not sufficient, for successful advocacy for infrastructure funding in the POM. In a range of risk and decisionmaking fields, several themes repeatedly arose: the need for high-level institutional buy-in, education of nontechnical personnel, collaboration and iteration to establish decisionmaking values and criteria, and the importance of developing a robust institutional decisionmaking environment and process. More compelling communication of mission risk is one of many needed elements.

In light of our conclusions, we offer several recommendations.

Recommendations

Assess the POM environment more deeply to determine the best way to implement the project scorecard approach. Of the three approaches we assessed, this seems to us the only viable one that can be implemented in the near term, to potentially improve on the status quo approach to presenting the POM request. We believe that the viability of this approach depends in part on the contents of the project tradespace, and in part on how the material is presented.

Continue to fully populate existing SMSs, and embrace and implement new ones as they are launched, with an eye toward informing a composite risk metric approach. The Air Force must do this to some degree anyway, in order to meet DoD's recent guidance.[4] But how far the Air Force goes in populating these SMSs, beyond the letter of the law, depends on the anticipated payoff. The Air Force can consult with the Army and Navy (as they appear to be further along in pursuing infrastructure performance and mission metrics, collecting data, and populating their SMSs) to see how their own investments have paid off.

Populating these systems should be a near-term priority, but some of the more powerful analyses—such as this risk assessment approach—will take time as data are gathered over the long term.

Make targeted assessments to determine when to use models to quantify mission outcome metrics. When done right, mission outcome metrics (and their supporting models) can provide especially compelling results, but they are narrow in scope and can also require significant effort. As a result, their application should be carefully calibrated to the desired outcomes. There may be cases where the mission outcome models could supplement the project scorecard approach in quantifying the mission contributions of some infrastructure projects.

Finally, **undertake high-level institutional action to educate stakeholders about the effects of infrastructure underfunding.** The civil engineering (CE) community greatly needs mission owners to help articulate the value of infrastructure in supporting Air Force missions and the dangers of infrastructure degradation, and mission owners need the CE community. The POM process itself is probably not a good forum for opening and educating minds about this admittedly complex topic, though there are a number of other possible avenues and forums for this education to take place. The CE community should consider as broad an approach to this as possible, as the obstacles are bigger than simply understanding the facts.

All of the steps we describe will require the Air Force and the CE community specifically to invest more time and effort. The challenge they confront is widespread, but no magic bullet exists. Other Air Force communities have also invested significant time and resources over many years in information systems and data to help inform requirements determination and POM

---

[4] A 2013 DoD memo mandated that (a) the Defense Components adopt a common process that incorporates the SMS, and (b) all real property assets shall have a validated Facility Condition Index by September 2017, including the provision that the condition data of each asset shall undergo a comprehensive validation on no less than a five-year cycle at minimum (DoD, 2013b).

advocacy. Given the criticality of infrastructure in the Air Force and the size of the annual investment (though presumably underfunded), it stands to reason that the Air Force must invest significant time and manpower in developing effective means to analyze and communicate the value of infrastructure funding to senior leaders and decisionmakers.

Acknowledgments

Numerous people both within and outside the Air Force provided valuable assistance to and support of our work. They are listed here with their rank and position as of the time of this research (fall 2015). We thank Maj Gen Theresa Carter, Air Force Civil Engineer, Deputy Chief of Staff, Logistics, Engineering, and Force Protection, for sponsoring this work. We also thank her staff for their time and support during this research.

In particular, we thank Edwin Oshiba (AF/A4C) and our action officer, Maj Greg Hoffman (AF/A4CI), for their fantastic support in providing access, information, and valuable insights.

At the Air Force Civil Engineering Center (AFCEC), we thank Lt Col Christopher Meeker and George Van Steenburg and their respective staffs for their time, insights, and a range of data and information that informed our analysis.

At Columbus Air Force Base, we thank Lt Col Robert Mozeleski, Capt Chad Fulgham, Don Young, Conny Boyd, and their staffs for providing their expertise and data to us for our case study.

Across the Air Force CE community, we thank the many people who took the time to provide detailed feedback on a draft version of this manuscript. We have tried to faithfully address your many comments and concerns.

At RAND, we thank Kristin Van Abel for gathering and synthesizing information on several key topics. We thank Gina Sandberg for her help in editing and formatting this manuscript. We thank Debra Knopman and Myron Hura for reviewing this work and helping to sharpen our analysis. We thank Jerry Sollinger for editing this report and helping to make it clear and concise.

Responsibility for the content of the document, analyses, and conclusions lies solely with the authors.

1. Introduction

> Assets exist to provide value to the organization and its stakeholders. Asset management does not focus on the asset itself, but on the value that the asset can provide to the organization. The value (which can be tangible or intangible, financial or non-financial) will be determined by the organization and its stakeholders, in accordance with the organizational objectives.
>
> —ISO 55000[1]

The success of any U.S. Air Force mission depends on the availability and performance of its support infrastructure. In some cases, the linkage between infrastructure and mission capability is clear: A closed runway directly affects sortie generation capability, for example. Most of the time, however, the connection is far less direct. While few would dispute that a poorly maintained runway increases aircraft wear and tear, eventually yielding greater fleet repair costs and reduced availability, such effects can be difficult to quantify or trace back to the underlying causes. As a result, the proper level of funding for infrastructure maintenance can be difficult to establish or defend, and the detrimental effects of chronic underfunding on mission capability and readiness may not become apparent for several years.[2] This analysis focuses on mission risk by exploring the relationship between Air Force infrastructure management and mission readiness and capability, with the goal of identifying data requirements and methodological approaches for illuminating and quantifying these links.

Infrastructure Degradation and the Outcomes of Inadequate Funding

All infrastructure assets deteriorate over time and with use. Infrastructure maintenance activities help mitigate that naturally occurring deterioration and maintain the infrastructure at an acceptable level. The Air Force often defines these maintenance activities as sustainment, restoration, and modernization (SRM). SRM includes activities ranging from preventive maintenance tasks, to periodic activities like regular roof replacement, to repairing damage of many kinds, to upgrading components or whole facilities to conform to recent standards.[3]

---

[1] International Organization for Standardization, International Standard 55000, *Asset Management—Overview, Principles and Terminology*, January 15, 2014, p. 3.

[2] In this report, we use the terms *readiness*, *mission capability*, and *mission performance* more or less interchangeably. *Readiness* often has very specific meanings, such as the financial accounts that underwrite training activities, or the actual readiness reporting systems and output metrics (e.g., C-ratings). Here, we use all these terms in a fairly generic sense of the ability to perform a given set of tasks or objectives.

[3] For more expansive definitions, see Air Force Instruction 32-1032, 2014. In commercial practice, this is usually referred to as maintenance, renovation, and reconstruction (MR&R) or maintenance and repair (M&R), though these three terms—SRM, MR&R, and M&R—are not completely overlapping.

There is no single way to define "enough" funding to sustain infrastructure, though practitioners often use wording such as "maintain the infrastructure in good working condition until the end of expected service life." This encapsulates the expectations of engineers who designed a facility, as well as end users who utilize it.

A notional depiction of how asset condition changes over time with and without normal maintenance activities is shown in Figure 1.1. In this example, the asset is assigned a condition rating at each point in time. Reduced infrastructure maintenance budgets yield reduced or delayed maintenance activity, resulting in, on average, a more degraded infrastructure condition, increased likelihood of major repairs, and, ultimately, reduced service life.

Figure 1.1. Effect of Maintenance and Repairs on Infrastructure Condition and Service Life

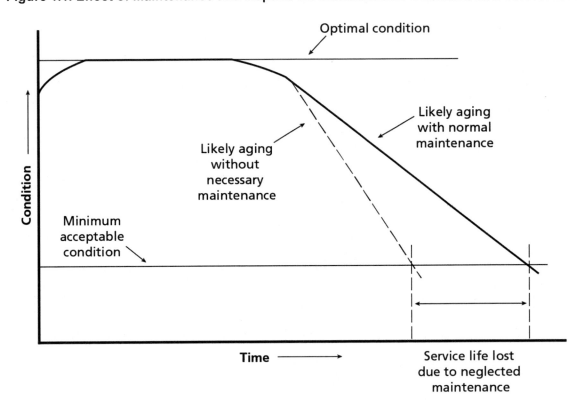

SOURCE: National Research Council, 1998.

Just as there is no "right" way to define enough funding, there is also no "right" way to quantify it for the purposes of budgeting. A 1990 National Research Council (NRC) report made the following recommendation, which became a standard for public facilities management:

> An appropriate budget allocation for routine M&R [maintenance and repair] for a substantial inventory of facilities will typically be in the range of 2 to 4 percent of the aggregate replacement value of those facilities (excluding land and major associated infrastructure). In the absence of specific information upon which to base the M&R budget, this funding level should be used as an absolute minimum value. Where neglect of maintenance has caused a backlog of needed repairs to

2

accumulate, spending must exceed this minimum level until the backlog has been eliminated. (National Research Council, 1990, p. xii)

In fact, there exist a range of methods to estimate sufficient infrastructure funding, including rough planning factors like the one above, which draw from broad commercial experience; more detailed parametric models, also based on commercial practice; and the cost of deferred maintenance and repair projects, driven by actual demands from engineers seeking to maintain the infrastructure. Therefore, when we use a term such as *underfunding*, it is not relative to some absolute standard, but one of a number of competing estimation methods and maintenance practices.[4] Often, underfunded infrastructure is explained in terms of its observed consequences, namely, deterioration.

Underfunded infrastructure budgets and the deterioration that follows are deep, widespread problems in the United States, both inside and outside DoD, and have been for some time. As early as 1988, the book *Fragile Foundations: A Report on America's Public Works* concluded that "unless we dramatically enhance the capacity and performance of the nation's public works, our own generation will forfeit its place in the American tradition of commitment to the future" (National Council on Public Works Improvement, 1988, p. 1). A 1989 article argued that "maintenance budgets are routinely starved by governments at all levels. Neglect, not age, is the root cause of most infrastructure failures in this country" (Regan, 1989). The 1990 NRC report quoted above found that "credible analyses indicate that we are systematically neglecting the maintenance of public facilities at all levels of government" (National Research Council, 1990, p. ix).

In 2003, the U.S. Government Accountability Office (GAO) designated federal real property as a "high-risk" topic for a number of reasons, among them deteriorating facilities (GAO, 2015). According to a 2008 special report commissioned by the American Society of Civil Engineers, "years of underfunding" have "allowed America's infrastructure to deteriorate" (Reid, 2008). Repeatedly (and as recently as 2013), the American Society of Civil Engineers' "Report Card for American Infrastructure" has given the United States dismal ratings (American Society of Civil Engineers, 2013). In 2013, GAO documented how roughly one quarter of bridges in the United States were "classified as deficient" and recommended that the federal government "review and evaluate funding mechanisms to align funding with performance," in this case bridge condition and serviceability (GAO, 2013).

This phenomenon is apparently now so common that a 2016 article in the *New Yorker* stated that "From the crumbling bridges of California to the overflowing sewage drains of Houston and

---

[4] We add to this ambiguity the way the U.S. Department of Defense (DoD) uses the word *requirement*. A now-outdated version of Joint Publication 1-02 defined a military requirement as "[a]n established need justifying the timely allocation of resources to achieve a capability to accomplish approved military objectives, missions, or tasks" (Joint Publication 1-02, 2005). Thus, the Air Force uses models like the ones described above to develop its infrastructure funding "requirements," but the requirements are simply statements of need to meet an intent, and those needs are derived from one among a number of competing methods for estimating infrastructure funding needs.

the rusting railroad tracks in the Northeast Corridor, decaying infrastructure is all around us, and the consequences are so familiar that we barely notice them" (Surowiecki, 2016).

Scott Gabriel Knowles (2014) argues that the extent of degradation of U.S. infrastructure is further evidenced by the damage from many recent natural disasters, including the failure of the New Orleans levee and dewatering system in Hurricane Katrina and the "Northeast blackout of 2003." Although addressing resilience to disasters as a specific concern is beyond the scope of this analysis, those in the Air Force and DoD concerned with mission assurance and critical infrastructure ought to take notice.

For DoD and the Air Force, degraded infrastructure can affect mission readiness in many ways. The future-year cost of infrastructure maintenance can grow, as deferred minor maintenance turns into expensive major repairs. These increased costs can reduce the funding available for other mission-related activities. Mission risk can increase, as critical degraded equipment and other infrastructure become more likely to fail or require unplanned outages. In extreme cases, poorly maintained infrastructure can give rise to civil code violations or environmental, health, and safety hazards. Poorly maintained infrastructure, particularly facilities associated with morale, welfare, and recreation, can reduce troop morale, health, and fitness, which will in turn degrade mission readiness in less easily quantifiable ways. We focus in this report on mission risk, but both risk and cost merit fuller descriptions here.

Cost Growth

A major theme of prior studies is that failures to perform necessary planned maintenance activities often lead to more costly unplanned repair activities. According to the "Law of Fives" heuristic, "if maintenance is not performed, then repairs equaling five times the maintenance costs are required" in later years (De Sitter, 1984). Even ignoring unplanned maintenance or the need for additional activities, the costs of planned maintenance activities increase as the conditions of facilities degrade. For example, one study estimated that the cost of a one-inch pavement overlay was $1.90 per lane-yard for a new pavement, but the cost increased to $19.90 per lane-yard for a pavement that had deteriorated into the worst state of disrepair (Durango and Madanat, 2002). When infrastructure maintenance budgets are underfunded and SRM activities are deferred, costs associated with maintaining, repairing, and using infrastructure grow.

User costs also increase as infrastructure condition degrades, in a variety of ways. For example, a study of bridge maintenance requirements in the United Kingdom found that costs to close or restrict usage of bridges in need of maintenance could dramatically overshadow the amount by which maintenance budgets were underfunded: "Underfunding would, in time, lead to bridges being closed or restricted while awaiting repair. The main effects would be road user delay costs of about £5.7 million [GBP] a year for each £1 million [GBP] of essential maintenance not undertaken" (Narasimhan and Wallbank, 1999). The same phenomena are found within DoD. Pavement in poor condition reduces fuel efficiency, increases the operating costs of vehicles using the pavement, increases accident risk, and can increase delay (Kuhn, 2011).

Mission Risk

Degraded assets can lead not only to increased future costs, but they can also make assets more susceptible to unplanned outages or catastrophic failure, thus compromising overall mission assurance. For example, in July 2014 in Los Angeles, a break at the juncture of two water lines sent more than 20 million gallons rushing through the University of California, Los Angeles (UCLA) campus, flooding underground parking structures, sport facilities, and several campus buildings. UCLA is seeking $13 million in damages from the city-owned utility, Los Angeles Department of Water and Power (LADWP), which owns and maintains the water lines that ruptured. It took approximately four hours for LADWP to completely shut off the lines. LADWP officials reported that "one of the lines that burst was badly corroded and the other had had five leaks before the rupture" (Gordon, 2015). Neither line was scheduled for replacement.

DoD defines mission assurance as

> a process to protect or ensure the continued function and resilience of capabilities and assets—including personnel, equipment, facilities, networks, information and information systems, infrastructure, and supply chains—critical to the performance of DoD mission essential functions in any operating environment or condition. (DoD, 2012)

According to DoD, mission assurance is achieved through the application of a "comprehensive risk management framework . . . based on mission-essential function and supporting asset prioritization" (DoD, 2012). The potential failure or inadequacy of critical infrastructure poses a risk to mission success that must be understood and managed.

Infrastructure-related mission risk cannot be managed (or communicated) without a clear understanding of how assets are related to mission requirements and which critical functions may be compromised when those assets operate at less than peak capability.

It is also possible to overfund infrastructure management; allocating infrastructure resources to activities that yield little benefit misuses resources that could be used more productively elsewhere. Assuring continued mission readiness at the minimal cost requires an understanding of, and the ability to articulate, the effects of each marginal dollar on the ability to carry out Air Force missions.

Why Infrastructure Maintenance and Repair Get Deferred

A fair question at this point is why, if infrastructure funding is so critical, is it so often given short shrift? For public institutions, some of the reasons often cited involve structural incentives: Officials get more credit for new infrastructure spending than for maintenance spending (Knowles, 2014); short terms in office mean that there are few electoral advantages in supporting projects that pay off years after officials have left office, and if officials cut maintenance spending, they may not be around when things go wrong (Knowles, 2014); budgeting and accounting processes create disincentives for cost-effective investments in maintenance and repair (NRC, 1998); and the distributed nature of decisionmaking about federal facilities investments can result in a lack of accountability for stewardship (NRC, 1998, 2004).

In the DoD and Air Force budgeting processes, infrastructure maintenance must compete against priorities that are, arguably, more compelling, more tangible, more immediate, and more familiar to many decisionmakers (e.g., weapon system modernization, training flying hours). Research on human judgments and decisionmaking shows that we often make simplifications to enable us to deal with complex information that lead to systematic and predictable errors, especially in the face of uncertainty.[5] These work against a cold, impartial assessment of the facts. Moreover, cognitive overload (which decisionmakers in the budgeting process must surely experience) can often push people toward their decisionmaking biases.[6] Thus, the structure and dynamics of the current system tend to work against adequately funding infrastructure maintenance.

Two more themes in the literature on decisionmaking, in particular decisionmaking about infrastructure maintenance, are the lack of credibility of those presenting budget requests and the lack of adequately educated decisionmakers. The first shortcoming is often linked to the lack of visibility and explanatory power of high-level cost models. The second (and related) shortcoming is a responsibility of infrastructure operators and managers.[7]

We raise these concerns because understanding the target audience and operating environment must help shape the Air Force's strategy to communicate the impact of infrastructure funding. It may not simply be a paucity of high-quality information that leads to underfunding decisions (though that is often true), but also common human behaviors and the existing institutional and decisionmaking environments. This report addresses the ways to develop and present better information in the Air Force's Program Objective Memorandum (POM) process, but history has shown that that alone may not be sufficient.

However, not all organizations fail to secure adequate infrastructure funding. A 2004 NRC report identified common best practices in infrastructure management in organizations that had

---

[5] For example, see Bazerman, 1998. People have been shown to misjudge probabilities in many ways, to bias assessments toward things that are familiar (representativeness heuristic), and to be unduly influenced by recent, memorable, or successful experiences (availability heuristic). People also sometimes show a tendency to over-discount the impact of future events (sometimes referred to as hyperbolic discounting). In another example (social facilitation), people exhibit the tendency to perform differently when in the presence of others than when alone: They perform better on simple or well-rehearsed tasks and worse on complex or new ones.

[6] In cognitive psychology, *cognitive load* refers to the total amount of mental effort being used in the working memory. Cognitive load theory was developed out of the study of problem solving by John Sweller in the late 1980s.

[7] One Air Force observer noted that, to some degree, the Air Force and the other services suffer from the inability to create a common nomenclature with fixed, universally used definitions in describing facilities and facility requirements. The ensuing confusion detracts from the Air Force's ability to educate the layperson as to the source and credibility of data. (Feedback provided by personnel from the Air Force Civil Engineering Center (AFCEC), provided on November 17, 2015.)

been successful in obtaining adequate funding for their maintenance and repair programs.[8] Among those practices were the following:

- Facilities are closely aligned with the organization's mission.
- Maintenance and repair investments are linked to the organization's product delivery or bottom line.
- The effects of failure are correlated with the organization's mission.
- Repair delay is correlated with sustainment cost.

The NRC recommended that organizations link maintenance and repair investments "to achievement of agencies' missions and other public policy objectives" (NRC, 2012), which is exactly what the Air Force seeks to do. But this and other reports do not actually lay out an approach to do so—in particular, how to synthesize and present analysis of mission impacts that decisionmakers find compelling. They often describe information systems, tools, and metrics that many organizations (including the Air Force) use—some of which we discuss in later chapters—but these do not by themselves articulate the impact of levels of funding on mission outcomes (especially in the outcome-relevant language that decisionmakers so crave).

Purpose of This Report

In fiscal year (FY) 2014, the Air Force Civil Engineer asked RAND Project AIR FORCE (PAF) to investigate how the Air Force might articulate the effects of SRM underfunding on readiness to ensure adequate funding to support these activities. Among the potential effects of underfunding infrastructure maintenance described above, we focus primarily on mission risk or impact. This project was originally scoped as a multiyear effort; this report represents the first year of research.

The broadly stated best practices in the previous section are exactly what the Air Force wishes to understand and apply. Unfortunately, we found no "playbook" from which the Air Force can simply tear a page. In this report, we describe and evaluate several divergent approaches to the problem, drawing on literature in a number of fields and the practices of other DoD services and government agencies.

We see this challenge as mainly comprising two elements. The first is risk analysis: how to gather and process data and information about infrastructure in a feasible, cost-effective way. This aims at developing and using information systems, employing engineers and end users to populate those systems, and developing analytic approaches for synthesizing those data in a meaningful way.

---

[8] NRC, 2004. We admit that most commercial firms have two advantages over the Air Force: They do not suffer the same structural incentives the government often does, and they can usually articulate the value of their infrastructure in terms of profits, something the government cannot do. But some of the techniques and strategies are potentially adaptable for DoD purposes.

The second task is about risk communication: choosing operationally relevant metrics that will resonate and compel, and organizing, winnowing, and presenting information to target the audience's absorptive capacity.

A 2004 NRC report summarized best practice approaches in infrastructure management as follows:

> Establish a framework of procedures, required information, and valuation criteria that aligns the goals, objectives, and values of their individual decision-making and operating groups to achieve the organization's overall mission; create an effective decision-making environment; and provide a basis for measuring and improving the outcomes of facilities investments. The components of the framework are understood and used by all leadership and management levels. . . . Best-practice organizations evaluate facilities investment proposals as mission enablers rather than solely as costs. (NRC, 2004)

That description encompasses several tasks that are beyond the scope of this report, including soliciting broad stakeholder input, educating less technically informed decisionmakers, and creating an overall environment of decisionmaking that favors wise investment. But we keep that picture in mind as we explore the more technical side of articulating the effects of infrastructure funding. In the final chapter of this report, we revisit this broader picture to put into context our recommendations and suggest a path forward for the Air Force.

Scope of This Report

Among the potential effects of underfunding infrastructure maintenance described above, we focus primarily on mission risk or impact. When discussing funding, we include the entire scope of SRM activities and therefore exclude categories such as family housing and military construction. We define SRM more fully in Chapter Two, but, in brief, sustainment includes routine tasks such as preventive maintenance, as well as cyclical tasks such as component replacement or repair; restoration restores components due to damage; and modernization improves the functionality of infrastructure. These are funding categories used to describe activities. We use those rather than descriptions of work classifications, which sometimes overlap non-uniquely with SRM categories.

Our discussion focuses on built infrastructure that is considered real property.[9] We generally exclude natural infrastructure and the equipment and personnel used to maintain and repair infrastructure assets. The approaches we explore are broadly applicable, but our examples focus on SRM.

---

[9] DoD defines real property categorization in Department of Defense Instruction 4165.03, 2015.

Organization of This Report

In Chapter Two, we discuss current Air Force infrastructure management practices, to set the stage for the rest of the report. In Chapter Three, we describe our analytic approach and discuss several broad analytic approaches and tools currently used in DoD and commercial practice. In Chapter Four, we describe a case study we performed to dig deeper into one of those approaches described in Chapter Three: building mathematical models of mission outcomes. In Chapter Five, we describe how the three approaches described in Chapter Three could be applied to the Air Force, and we weigh various advantages and disadvantages (including what is involved in implementing them). In Chapter Six, we provide overall conclusions and make some recommendations for implementation by the Air Force.

2. Current Infrastructure Management Practices

In this chapter, we describe current Air Force infrastructure management practices. First we describe the scope and definitions of infrastructure management, then the processes by which POM funding requests are developed and how that funding is allocated.

Air Force Infrastructure Management

Each year, the Air Force allocates more than $10 billion to provide, operate, and maintain installations[1] valued at nearly $275 billion (DoD, 2013a, p. 7).[2] The Air Force manages a broad range of infrastructure types, from runways, vertical structures, and communications infrastructure to water and power production and distribution. These infrastructure assets support a set of missions, which include day-to-day peacetime training and institutional support (e.g., aircraft training operations, laboratories, test facilities), home-station employ-in-place activities (e.g., nuclear, space, cyber, and remotely piloted aircraft operations), and combat fight-in-place garrison operations, especially at overseas locations.

The Air Force Civil Engineer and base civil engineers (BCEs)[3] sustain the array of infrastructure assets and systems using a range of small- and large-scale activities, much of which fall under the umbrella of SRM.[4] These activities and the budgets associated with them are the focus of this analysis. In this report, we define each SRM component as the Air Force does, as follows:

Sustainment activities are categorized as either sustainment maintenance or sustainment repair. *Sustainment maintenance* is defined as

> work to maintain the inventory of real property assets through its expected service life. It includes regularly scheduled adjustments and inspections, and preventative maintenance tasks. Maintenance is routinely completed through the Recurring Work Program and Direct Scheduled Work Program. There may be

---

[1] Secretary for the Air Force, Deputy Assistant Secretary for Budget, 2014. This amount includes installation support, SRM, and military construction at all locations worldwide.

[2] While the spending figure is 4 percent of the total value, this should not be construed to mean that the Air Force is hitting the target of spending 2–4 percent of replacement value per year on infrastructure SRM. The $10 billion figure includes all sources of funding.

[3] The BCE is the lead civil engineer at an Air Force installation and could be either military or civilian. He or she is also the lead Emergency Manager, responsible for ensuring that the base's command and control (C2) system is in place.

[4] Facility operation is another large category. It includes such activities as custodial services, grounds services, waste disposal (the three of which are often called the "Big Three"), and the provision of central utilities. These are usually provided by contract support and comprise a significant share of the infrastructure budget.

times when a contract effort is necessary to complete maintenance work. (Air Force Instruction 32-1032, 2014, p. 54)

Sustainment repair is defined as

> scheduled repair activities to maintain the inventory of real property assets enabling them to reach their expected service life. It includes emergency response and service calls for minor repairs. It also includes major repairs or replacement of facility components (usually accomplished by contract) that are expected to occur periodically throughout the life cycle of facilities, and any repairs to inadequately-sustained components. This work includes regular roof replacement, refinishing of wall surfaces, repairing and replacement of heating and cooling systems, replacing tile and carpeting and similar types of work. Timing of the work (within or post life cycle) isn't the determining factor between sustainment and R&M—the purpose of the work is the primary factor. Life cycle repairs accomplished post-expected life cycle (e.g., deferred, delayed, neglected) are still sustainment repairs. (Air Force Instruction 32-1032, 2014)

Restoration "includes repair and replacement work to restore facilities collaterally damaged due to inadequately sustained components, natural disaster, fire, accident or other causes." *Modernization* "includes alteration of facilities or components solely to implement new or higher standards (including regulatory changes and code compliance), or to accommodate new activities" (Air Force Instruction 32-1032, 2014). (Demolition and consolidation are separate funding categories.)

Current Air Force Infrastructure Funding Processes

We now walk through the Air Force's infrastructure funding and management processes to understand how funding affects infrastructure sustainment and maintenance activities.

Requirements Are Developed Annually

We start at the lower left of Figure 2.1, with requirements development. Each year, each base develops a Base Comprehensive Asset Management Plan (BCAMP). An asset management plan—a widely used tool in infrastructure management—is a structured, standardized approach to manage infrastructure assets from a holistic portfolio perspective. BCAMPs compare current and future requirements with available assets and produce plans to sustain, repair, or upgrade current assets; dispose of surplus; and construct new assets as needed. They seek to meet stated requirements at minimum life-cycle cost. These plans are, naturally, multiyear. Requirements include organic manpower (military and civilian) to sustain and repair infrastructure assets equipment and supplies, as well as provide oversight, training, etc.

Figure 2.1. Processes for POM Input and Resource Allocation

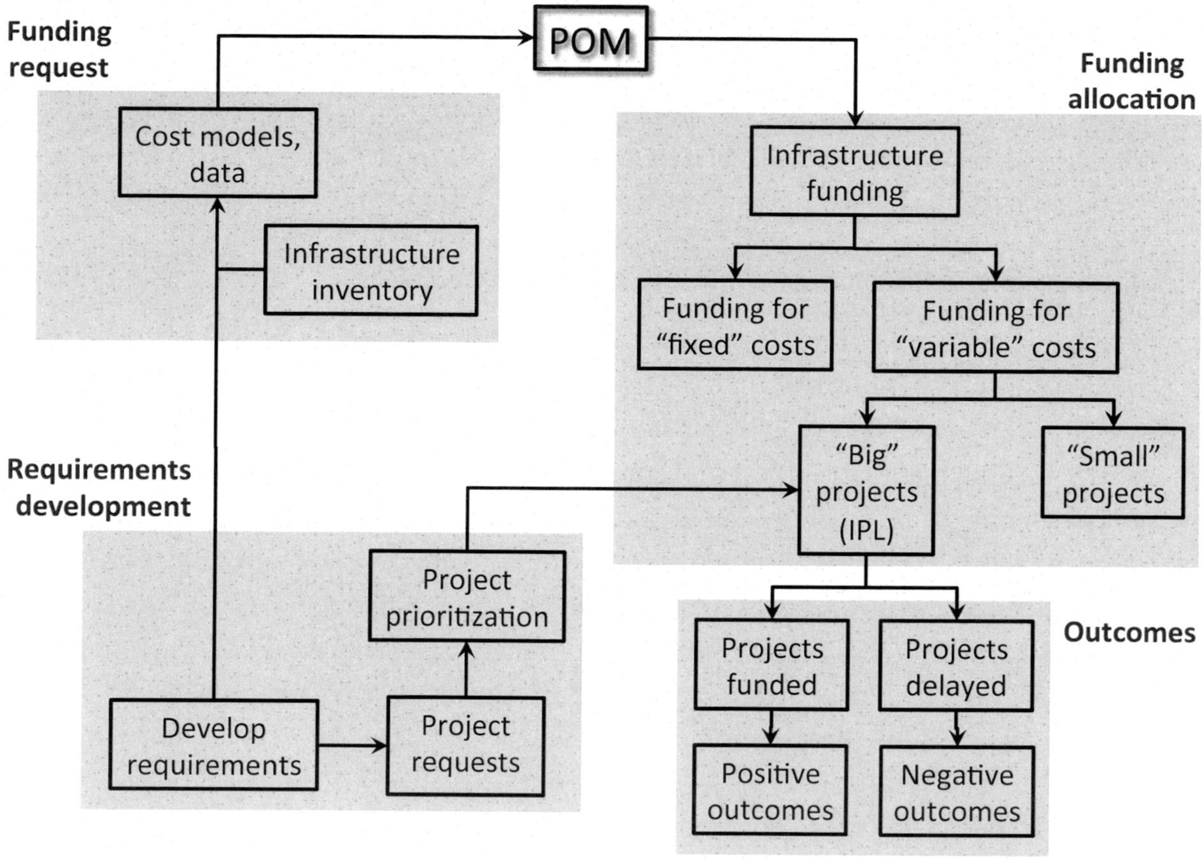

NOTE: IPL = integrated priority list.

Requirements also include the need for larger construction activities called *projects*. A project is defined as "any maintenance, repair, construction, or combination of the three performed on or in a facility necessary to produce a complete and usable facility or improvement to a facility . . . regardless of dollar amount or execution strategy."[5] Projects range widely in size and scope. These projects include activities that are beyond the scope of regular sustainment activities, usually because an asset is at a critical point—that is, it has either been neglected for a long time, there is an impending failure, or the repair needs of the infrastructure asset exceed the base's organic repair capabilities. Examples of such projects are replacing the roof on a building, repairing a section of runway, replacing or repairing a chiller or a boiler, or major repairs to a wastewater line. Single projects can consist of several requirements bundled together. These

---

[5] Combining requirements in multiple facilities for a single contract or task order is an execution strategy, which could include multiple "projects" (a project is specific to a single facility). However, there may also be multiple projects within a single facility if each individual project fulfills an individual requirement that produces a complete and usable facility or improvement to a facility (or component of a facility) and is independent and not interrelated with other requirements within the facility (Air Force Instruction 32-1032, 2014).

could range in cost from thousands to many millions of dollars. Each BCAMP produces a single list of near-term projects that need to be completed to support base needs.

BCAMPs are assembled by each major command (MAJCOM) to produce a MAJCOM Comprehensive Asset Management Plans (CAMP), and MAJCOM CAMPs are assembled across the Air Force to produce an Air Force CAMP (AFCAMP). The project lists produced by each base are assembled into a single Air Force project list. We discuss the processes of prioritizing the projects and allocating resources to them later in this section.

Infrastructure Funding Request Uses High-Level Cost Models

On the upper left side of Figure 2.1, the core of POM input is a series of cost models and data analyses, informed by base-level requirements determination and enterprise-level information inputs about the current Air Force infrastructure inventory. The key model used in this process is the Facility Sustainment Model (FSM), a parametric model that estimates the amount of sustainment funding necessary to adequately sustain infrastructure of various types.[6] FSM uses cost planning factors to maintain infrastructure consistent with commercial practices through their expected lives. It applies annual spending factors (e.g., sustainment cost per square foot) based on physical dimensions for different classes of infrastructure. These tools marry these planning factors with the actual inventory of Air Force assets in each category. The tools then arrive at a total amount of sustainment funding needed to sustain these assets. Note that the project lists produced at base level do not directly inform the POM; funding estimates for those activities (e.g., SRM) are produced by these high-level cost models.

The various models and planning factors used result in an overall funding request that can be described broadly in terms of the categories of activities it supports (e.g., preventive maintenance, emergency repairs), but cannot tie those activities to outcomes when underfunded. These cost inputs are then reviewed by the Air Force Corporate Structure (AFCS), starting with the installations panel, then going to the Air Force Group, Air Force Board, Air Force Council, and finally the Chief and Secretary of the Air Force (Air Force Instruction 16-501, 2006). Each year in the POM process, funding over the entire Future Years Defense Plan (FYDP) is determined, which has multiyear ramifications. Because the FYDP funding is revisited (though not completely from the ground up) each year, it provides a new opportunity to secure needed funding.

---

[6] Each year, the Office of the Secretary of Defense contracts with a company to update the FSM (according to its website, RK Solutions developed the original FSM for DoD). FSM is based on industry standards that define costs to keep an inventory of facilities functional through their expected service lives. For more information, see Whitestone Research and Jacobs Facilities Engineering, 2001. In addition to a sustainment model, there are associated recapitalization and operating cost models. See Lufkin, Desai, and Janke, 2005.

Funding Is Allocated to Several Infrastructure Categories

On the right side of Figure 2.1, a level of infrastructure funding is appropriated by Congress, then allocated by the Office of the Secretary of Defense and the Air Force. Ultimately, that infrastructure funding is divided among a range of smaller categories and activities. Here, we divided this into three broad categories. Some portion goes to what we refer to as fixed costs. A large part of these fixed costs is allocated directly to the bases. The BCEs use that funding for day-to-day infrastructure sustainment. BCEs accomplish routine maintenance tasks as specified by manufacturers and complete urgent and emergency tasks according to on-base customer requests, all within their given organic capacity and the funding provided. This base-level maintenance underpins the health of the Air Force's infrastructure. To the degree that they have the capacity (with organic military or civilian manpower, plus some equipment and supplies), the BCEs also take on larger tasks.

The rest of the funding goes to projects. Each base receives some discretionary funding for contractors to take on projects beyond their organic capacity but below some dollar threshold. In this report, the SRM projects we speak to mostly fall within the portion of sustainment repair that includes major repairs or replacement of facility components and restoration.

Larger projects (i.e., above some dollar threshold) are handled separately. Funding for large projects is allocated through a formal, enterprise-level prioritization model, as part of the AFCAMP project prioritization process. In this process, each project competes against all the other projects in that funding cycle.[7] We now describe that process.

Projects Are Prioritized to Produce an Integrated Priority List

The BCAMPs developed at base level include requests for infrastructure projects, including the scope of the project (e.g., roof replacement for first fighter squadron operations building), its projected cost, and data inputs to feed the prioritization model.[8] Then, during the year of execution, AFIMSC creates a single integrated priority list (IPL) using an equation that weights those data elements provided by BCEs to assign a single score for each project. The project score

---

[7] Before the realignment of sustainment funding from the MAJCOMs to the Air Force Installation Mission and Support Center (AFIMSC) starting in FY 2016, there was a $5 million lower limit for consideration of sustainment requirements for placement on the IPL. That has now been removed. Bases and MAJCOMs may submit all projects that could be funded using the centralized funding that AFCEC manages for the AFCAMP program, regardless of cost, and all must have a base priority assigned. There is no longer a lower cost threshold (U.S. Air Force, 2015).

Now, once all the project submissions are received, AFIMSC will determine a minimum dollar threshold for IPL projects. Those above the threshold will go into the IPL process; those below will revert back to base level for accomplishment by organic personnel or contract support. It is estimated that this threshold will be somewhere between $100,000 and $300,000, with work order supply funds increased slightly to compensate for increased base-level workload. (Source: Email conversation with personnel from HQ AFIMSC/IZBS on March 22, 2016.)

[8] Until recently, each category of infrastructure project (e.g., sustainment, demolition, dorms, energy) had been a separate program that was prioritized and executed according to its own rules. Under centralization, a large majority of the civil engineering (CE) programs have been integrated and are now prioritized on a single list. (Source: Email conversation with personnel from HQ AFIMSC/IZBS on March 22, 2016.)

comprises three terms: the probability of failure (PoF), a value from 1 to 100; the consequence of failure (CoF), a value from 1 to 100; and, if applicable, the cost savings, a value of 1 to 10.[9]

Allocation Decisions Lead to Positive and Negative Outcomes

The funding gained in the POM process is applied to the prioritized project list ("big" projects on the right side of Figure 2.1) from highest priority to lowest, and projects are funded until the money runs out. For the funded projects, affected infrastructure assets are in some way improved, and thus they experience positive outcomes.

Projects that are "below the line" are delayed until the next funding cycle (or until other sources of funding can be secured—e.g., emergency funding if the project justifies it). Infrastructure assets whose projects were delayed experience further degradation and negative outcomes (e.g., higher costs, lower performance).

If a project is not funded in the current funding cycle, the local base-level CE unit would be responsible for maintaining the status quo, including addressing any damage incurred until the asset or component could be repaired or replaced. Arguably, the longer the project waited, the more local organic time and resources would be devoted to maintaining the asset in serviceable condition.

Funding Allocation Is Evident in Infrastructure Spending Patterns

Our analysis of recent spending data from the Commanders' Resource Integration System (CRIS) supports the narrative explanation above.[10] We found that within the whole of infrastructure operations and support spending, a significant portion of that spending—about $1.1 billion per year—is fixed in relation to the total spending. The activities that stayed more or less constant regardless of total spending fluctuations included, but were not limited to, organic base-level capacity (e.g., civilian pay, equipment, supplies),[11] utilities, and safety-related activities, such as elevator certification.

In addition to the fixed costs, a small set of categories varied directly with the total level of infrastructure spending: sustainment projects, repair and maintenance (R&M) projects, and Simplified Acquisition of Base Engineer Requirements (SABER).[12] These categories, most of which go to contractor support, made up nearly all the spending fluctuations in this time period.

---

[9] Projects that have nothing to do with risk but still present savings opportunities are considered as savings-only projects and are evaluated only on savings-to-investment ratio (SIR) and subjectively placed in the IPL according to Air Force goals (U.S. Air Force, 2015).

[10] Accounting data provided by AF/A7C from Commanders' Resource Integration System (CRIS), FYs 2004–2012.

[11] Manpower levels for active-duty military civil engineers (most of whom directly support SRM activities while at home station) were also relatively flat during this time period, but military personnel are funded through a different appropriation than operations and support and thus were not included in this data set.

[12] A SABER contract's main purpose is to expedite contract award of civil engineer requirements through the issuance of individual delivery orders. This is an instrument that is often used at base level to fund projects beyond the scope of their organic capacity.

These variable costs ranged from about $1 billion to $2 billion per year during this time frame, composing between one-half and two-thirds of total infrastructure spending.

This means that within certain bounds, organic base-level CE organizations are mostly unaffected by near- and medium-term funding fluctuations (i.e., one FYDP or so). When infrastructure funding is tight, projects are postponed, and when funding is eventually secured, those postponed projects can then move forward, or at least re-compete with new projects that may have arisen in the meantime. This means that when the AFCS is considering a total level of infrastructure funding, what it is really debating (in effect) is how much money will go to projects, and thus which infrastructure assets will have their projects delayed until the next funding cycle, and will therefore be allowed to degrade in the meantime.

One challenging aspect of utilizing the POM process (as part of the larger Planning, Programming, Budgeting, and Execution System [PPBES]) to allocate funding for infrastructure is that infrastructure assets have design lifetimes of many years or decades, while funding is reconsidered every year and can fluctuate significantly. Many projects can be shifted or postponed without much mission impact or cost growth, but they must be done eventually. If funding truly needed to sustain infrastructure is diverted to other, presumably more pressing needs, even more funding has to be secured in future years, a difficult prospect in the zero-sum game of DoD budgeting.

In the next section, we describe in more detail the Air Force's project prioritization scheme and resulting project list.

Current Infrastructure Project Prioritization Model

The current prioritization model for built infrastructure has three overarching elements: probability of failure (PoF), consequence of failure (CoF), and cost savings.[13] The equation below weights each of these data elements accordingly to arrive at a single project value from 0 to 210. (In the equation, CI = condition index, MDI = mission dependency index, and SIR = savings-to-investment ratio.)

Project Score = ([100 − CI] × 2) + (0.6 x MDI + 0.4 × MAJCOM Priority Points) + (SIR x 10)

Probability of failure is represented by the CI, a value from 0 to 100. The three major asset types—facilities, transportation and pavements, and utilities—draw their CIs from different sources. The project CI is a cost-weighted average of CIs for all elements (e.g., systems, sections, components) in the project scope:

- *Facilities*. This is a value from 1 (failing) to 100 (excellent) reflecting the condition of the facilities asset. The value is generated by the BUILDER[TM] Sustainment Management

---

[13] This section draws heavily from U.S. Air Force, 2015. This equation has only condition index contributing to PoF, but the text makes allowances for functionality degradation. We explain this more in Chapter Three.

System (SMS)[14] and is based on life-cycle projections, actual facility condition, and functionality assessments, including but not limited to safety, Americans with Disabilities Act compliance, fire safety deficiencies codes, and space/capacity assessments.

- *Transportation and Pavements.* This is a value from 1 (failing) to 100 (excellent) reflecting the condition of transportation and pavement assets. The value is based on condition, foreign object debris potential, skid potential, and the structural index of the asset.
- *Utilities.* This is a value from 1 to 100 reflecting the remaining service life obtained from standard tables and the performance of the utility asset (e.g., documented breaks or outages).[15]

We note that the CI for facilities in particular utilizes a broader slate of metrics than some traditional CIs. Assessments of functionality target inherent characteristics of a facility that may be driven by safety, or by the mission activity housed within. A facility can be in good condition (in the traditional sense) but not support the functions required of it by its users. There is also an allowance in the PoF term for projects that include facilities experiencing a change in function or personnel consolidation. We discuss this in more detail in Chapter Three.

Consequence of failure comprises two criteria: MDI and MAJCOM priority.

- *Standardized MDI.* This value from 0 to 100 is intended to reflect mission dependence and importance. It is currently assigned based on the infrastructure category of the asset (e.g., a library scores 39, an aircraft maintenance hangar 70, and a control tower 90), with a process to adjust upward or downward based on location-specific mission information.[16] We discuss MDI more in Chapter Three.
- *MAJCOM Priority Points.* This value from 1 to 100 is weighted by plant replacement value (PRV).[17] For example, priority #1 receives 100 points. Points decrease

---

[14] The SMS is a suite of web-based software applications developed by the Engineer Research and Development Center's (ERDC's) Construction Engineering Research Laboratory (CERL) to help facility engineers, technicians, and managers decide when, where, and how to best maintain the built environment. The SMS modules include BUILDER™ and ROOFER™ for assessing building conditions, PAVER™ for pavements, and RAILER™ for railroad infrastructure. ERDC-CERL is currently developing FUELER, which will be an additional module to support the inspection of fuel storage and distribution facilities. Modules for other facility types (utilities, structures, etc.) are under various phases of investigation and development. The SMS suite provides an asset management solution to repeated GAO criticisms of past DoD facility management practices.

A 2013 DoD memo mandated that (a) the Defense Components adopt a common process that incorporates the SMS, and (b) all real property assets shall have a validated Facility Condition Index (FCI) by September 2017, including the provision that the condition data of each asset shall undergo a comprehensive validation on no less than a five-year cycle at minimum (DoD, 2013b).

[15] This value is generated from GeoBase Spatial Data Standards for Facilities, Infrastructure, and Environment.

[16] A Category Code (CATCODE) is a five or six-digit code used by DoD that represents a specific type of facility. CATCODEs are assigned in accordance with Department of Defense Instruction 4165.03, 2015.

[17] Air Force personnel brought to our attention that the Air Force uses multiple definitions of *plant replacement value* in its various facility activities. The primary difference in the definitions as used is the degree to which depreciation of a fixed asset is considered. The definition is not consistent. The FSM cost tool referenced earlier uses *plant replacement value* according to a common commercial definition, "Cost to replace the current facility as is in like or new condition," because that concept drives the commercial cost factors integral to the tool. Feedback provided by personnel from AFCEC-CP, provided on November 17, 2015.

proportionally based on the number of MAJCOM priorities. Currently, priorities are allocated to MAJCOMs based on their portion of total infrastructure PRV.

Potential cost savings of the project are based on the SIR, the ratio of project life-cycle savings to the total project cost (resulting in a value between 0 and 1), multiplied by 10. The sum of these three terms can thus range from 0 to 210.

This prioritization model (including its criteria and weights) was developed by AFCEC, socialized through the CE corporate process, beta-tested to demonstrate its results, and then implemented in the actual AFCAMP process.[18] This kind of prioritization model is common in multi-criteria decisionmaking. We discuss this type of decisionmaking tool more in Chapter Three.

For the sake of discussion, we developed a notional project list using an actual MAJCOM IPL worksheet (in Microsoft Excel) provided to us by AFCEC. We have restructured and edited it for the sake of legibility. Table 2.1 shows this notional project list with 34 projects (19 projects that we generated are not shown).

---

[18] Email conversation with personnel from AFIMSC/IZBS on May 2, 2016.

Table 2.1. Notional Project List

Project Title	Project cost ($M)	PoF Score	CoF Score	Savings Score	Total Score	Cumulative funding ($M)	
RAPCON CEN	27	38	95	3	136	500	<--19 projects not shown
POL OFFICE	24	60	71	5	136	524	
ACFT COR CTL	30	58	73	3	134	554	
PMEL MAINT LAB	33	66	59	6	131	587	
FLY TNG BLDG	26	70	58	2	130	613	
ACFT SVC EQPT (AGE)	9	36	85	6	127	622	
FLY TNG BLDG	47	48	74	5	127	669	
FABRICATION SHP	4	46	77	3	126	673	
THRIFT SHOP	4	64	53	8	125	677	
MAINTENANCE HANGAR	20	66	50	9	125	697	70% = $700M
PRIME BEEF WHSE	8	60	55	10	125	705	
HG MAINT HANGAR 7	11	48	62	9	119	716	
FLT SIM BLDG	8	56	48	9	113	724	
OFCR MESS FCLTY	39	64	43	3	110	763	80% = $800M
BASE MAIN GYM	50	24	83	2	109	813	
ALTITUDE CHAMBR	12	22	80	6	108	825	
JET ENG RPR FAC	28	46	59	1	106	853	
COMM BLDG	7	44	56	1	101	860	
BSE SUPPLY BLDG	21	40	47	8	95	881	
SF WAREHOUSE	10	32	55	5	92	891	
ENG TST CELL	7	8	74	9	91	898	
BE CV STORAGE	4	46	35	9	90	902	90% = $900M
WASH RACK - HANGAR	22	4	80	4	88	924	
JET ENGINE SHOP	9	42	45	1	88	933	
ACFT MAINT HANGAR	27	22	58	6	86	960	
FIRE CRASH RESCUE STATION	27	26	51	8	85	987	Request=$1,000M
YOUTH CENTER	21	38	47	0	85	1,008	
MAIN POL HQ	15	38	37	7	82	1,023	
POLICE STATION	26	28	41	5	74	1,049	
ACFT MAINT BLDG-HANGAR	33	18	47	4	69	1,082	
H/SHP AUTOMOTIV	20	4	54	10	68	1,102	
EDUCATION CENTR	16	20	44	3	67	1,118	
FAMILY SUP CEN	8	2	58	6	66	1,126	
ACFT CORR CTL	13	8	55	3	66	1,139	

We populated this table by selecting the names of building types from a BUILDER data set provided by AFCEC for Columbus Air Force Base (AFB). The MDIs listed are from the BUILDER data set. Otherwise, the other inputs—cost, condition index, MAJCOM priority, and SIR—are randomly selected, for illustrative purposes only. In the table, the white cells are the project name and cost ($1–$50 million), and light gray cells are calculations for the three major components of the utility function: PoF, CoF, and savings (the others are suppressed). The resulting total score is in the penultimate column, and the cumulative cost in the rightmost column (both in dark gray). We show only the list of projects with a cumulative cost from $500 million to $1 billion. To the right of the project list, we show cumulative funding levels for a notional 100 percent request of $1 billion, and at 90 percent, 80 percent, and 70 percent of that request.

In the BCAMP requirements-gathering process, base-level users from each MAJCOM populate supporting worksheets to generate final worksheets like this to be entered in a data system for download and synthesis by AFIMSC. For the remainder of the report, we will often refer to this as our notional project list. We note that the BCAMP process is a broader multiyear requirements-gathering process; only the most urgent (and larger) needs are translated into projects, a process that requires some additional investment of time and manpower to specify needs in enough detail to provide a reasonable cost estimate.

In the year of execution, a total IPL funding level is determined, which is then allocated against projects that are to be executed in the following year, from top to bottom on the IPL. In Table 2.1, we draw lines at 10 percent increments below the requested $1 billion, and we suggest that the band between each two funding numbers (and the projects in that space) represents the project tradespace. For example, if the AFCS is considering a figure of at least $800 million, the projects above the line (those in the first 14 rows of the table) are not even up for debate. They more or less automatically get funded via their priority. But if the AFCS is deciding between a funding level of $800 million and $900 million, then the seven projects listed between $800 million line and the $900 million line in Table 2.1 make up the project tradespace.

In the current process, the project list is not used in the POM deliberations. But the deliberations between one funding level and another are *in effect* deferring the projects inside that tradespace. In Chapter Three, a method for incorporating the project list into the POM deliberations to inject more concreteness into the process.

The question (posed to us) is, if a funding level of $1 billion is being considered, what is the impact of delaying the projects in the project tradespace? What are the risks of not funding those in the tradespace, and what are the rewards of those that get funded? We take up that question in the next chapter.

3. Three Approaches to Linking Infrastructure to Mission

In this chapter, we survey three broad approaches that are potentially useful for linking infrastructure to mission: a project scorecard approach, an approach based on mission outcome metrics, and an approach based on composite risk metrics. First, we explain our own methodology for identifying and assessing these three approaches. Then we briefly describe each analytic approach.

This Project's Methodology

Our methodology has two steps. First, we identified competing analytic approaches that have been used for analyzing and presenting mission risk (whether specifically for linking infrastructure to mission or not). To do so, we reviewed relevant literature from academia, commercial practices and case studies, and DoD policies and practices. We also discussed a range of approaches with Air Force and other service personnel involved in engineering and/or POM deliberations. The results of this step are presented in this chapter.

Second, we assessed each of these options along several lines. For all three cases, we assessed the option's strengths and weaknesses, including relative costliness to implement, in terms of time, manpower, and institutional energy (not actual investment cost to implement or recurring costs to maintain). We also explored ways to mitigate the weaknesses of each approach to make it most useful in the Air Force context. For one of the approaches we identified, we conducted a case study, in order to flesh out data requirements and assess the current availability of data to support the approach. We describe this case study in Chapter Four.

The rest of this chapter describes the first step, identifying and developing options.

Table 3.1 summarizes these approaches: the nature of their main output and the steps to produce them.

Table 3.1. Summary of the Output and Steps of Three Alternative Approaches

Attribute	Project Scorecard	Mission Outcome Metrics	Composite Risk Metrics
Output	• List of projects up for deliberation with affected missions	• Quantification of future mission performance based on funding levels	• Expected composite risk ratings based on levels of funding
Steps	• Develop and prioritize project list from base-level inputs • Develop project tradespace (groups of projects) based on levels of funding • Apply mission areas to projects in tradespace • Mission owners advocate for projects in their purview	• Develop mission metrics for a single mission • Link infrastructure assets to mission outputs with logic model • Develop mathematical model quantifying relationships between key assets and mission outputs • Impact of projects under consideration quantified in terms of mission metrics	• Assess mission performance (i.e., condition, function) of infrastructure assets • Develop and apply mission performance thresholds for infrastructure assets • Translate mission performance to risk metrics • Develop composite risk profiles based on levels of funding

Scorecard Analysis

A scorecard is simply a table of options (usually in rows) with various criteria (usually in columns) and some scoring of each option along each criteria.[1] The notional project list in Figure 2.1 could be considered an example of a scorecard.

In fact, scorecards are one element within a larger field of decision analysis called multi-criteria decision analysis (MCDA).[2] For context, we describe MCDA and some of its key elements briefly.

Multi-Criteria Decision Analysis

MCDA is both an approach and a set of techniques, with the goal of providing an overall ordering of options, from the most preferred to the least preferred option.[3] MCDA is a way of looking at complex problems that comprise multiple objectives, of breaking the problem into more manageable pieces to allow data and judgments to be brought to bear on the pieces, and then of reassembling the pieces to present a coherent overall picture to decisionmakers. The purpose is to serve as an aid to thinking and decisionmaking, but not to make the decision.[4]

---

[1] This is not to be confused with the "balanced scorecard" approach espoused by Kaplan and Norton. See their 1992 *Harvard Business Review* article (Kaplan and Norton, 1992) or their popular book, *The Balanced Scorecard: Translating Strategy into Action* (Kaplan and Norton, 1996).

[2] This is also sometimes called multi-criteria decision-making (MCDM) or multi-objective decision analysis (MODA). The term multi-criteria analysis (MCA) is sometimes used to refer to a somewhat broader set of approaches that includes MCDA and others. MCA is most commonly used in the Commonwealth.

[3] This section draws explanations of MCDA liberally from United Kingdom Department for Communities and Local Government, 2009.

[4] The first complete exposition of MCDA was given by Keeney and Raiffa, whose 1976 book, *Decisions with Multiple Objectives: Preferences and Value Tradeoffs*, is still useful today. They built on decision theory, which for

A key feature of MCDA is its emphasis on the judgment of the decisionmaking team in establishing objectives and criteria, estimating relative importance weights, and, to some extent, estimating the contribution of each option to each performance criterion. The main role of the techniques is to deal with the difficulties that human decisionmakers have been shown to have in handling large amounts of complex information in a consistent way.

Applying MCDA consists of a multistep process that generally enlists a range of stakeholders and elicits institutional values to guide decisionmaking. It has been significantly developed into a large set of analytic techniques and has been applied to a number of public policy decisions, including infrastructure resource allocation. The project prioritization model reflects a fairly simple MCDA approach. And what the AFCS does in the POM process is, of course, multi-criteria decisionmaking, though fairly informal.

Value Tree

Ultimately, the scorecard is a vehicle for communicating several (or many) criteria to a decisionmaker to weigh. Selecting the right criteria is the key to effectively weighing the trade-offs, so the process of deriving those criteria is an important step. Likewise, it is common practice to group the criteria, for several reasons: (a) to help check whether the set of criteria selected is appropriate to the problem; (b) to ease the process of calculating criteria weights in large MCDA applications;[5] and (c) to facilitate the emergence of higher-level views of the issues, particularly how the options realize trade-offs between key objectives (United Kingdom Department for Communities and Local Government, 2009).

The organized set of criteria is sometimes called a value tree because of its shape. Figure 3.1 shows an example of a value tree expressing the criteria from the current project prioritization model. The top level shows the overall project score, the second shows the three major elements, and the bottom row shows the sub-elements. In fact, the bottom level of sub-elements could be further broken out.

For example, functionality is assessed with several elements, including safety and space/capacity assessments. Likewise, MDI is a composite metric that assesses how critical an asset is to a mission by evaluating how interruptible and replaceable it is. Many organizations have value trees similar to the Air Force's, some with several layers of criteria.

---

most people is associated with decision trees, modeling of uncertainty, and the expected utility rule. By extending decision theory to accommodate multi-attributed consequences, Keeney and Raiffa provided a theoretically sound integration of the uncertainty associated with future consequences and the multiple objectives those consequences realize. Raiffa espoused many of the concepts in that book while at RAND in the 1960s (Raiffa, 1969).

[5] In large applications, it can sometimes be helpful to assess weights first within groups of related criteria and then between groups of criteria.

Figure 3.1. Value Tree of Air Force Project Prioritization Model

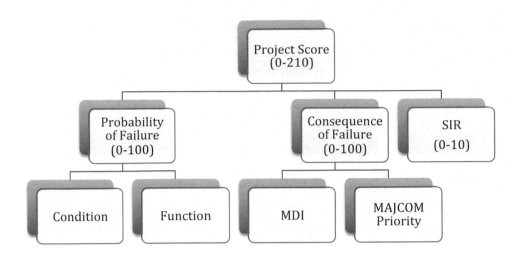

SOURCE: Authors' analysis of U.S. Air Force, 2015.

The value tree presents a range of possible inputs to a scorecard, but what goes into the scorecard must be decided. In the notional project list in Chapter Two, we showed the three main terms in the prioritization model and the final score. But, is it best for a decisionmaker to view only the top-level of three metrics, i.e., PoF, CoF, and SIR, as we showed in our notional project in Figure 2.1? Or the next level, with four? Or the next? Less information in a coarser form is easier to digest, but probably harder to intuit; incorporating information from lower levels means more concrete information that may be easier to intuit, but results in a higher volume of information to digest.

The Scorecard

A scorecard (also called a performance matrix, or consequence table, in some fields of decision analysis) is a standard feature of MCDA. According to Hillestad and Davis (1998),

> Historically, the scorecard, sometimes called a stoplight chart, has been an important method of presenting the results of a policy analysis to decisionmakers. . . . The two-dimensional display permits a quick view of how all of the policy options fare across the measures of interest. This is particularly important because good policy analysis recognizes that policymakers must bring to bear a number of value judgments and constraints that cannot and probably should not be buried in technical analyses done by their staffs. That is, there should be a separation between what can be accomplished "technically" and what must be assessed by the decisionmakers themselves. The scorecard approach permits this. For example, some of the columns may show "technical" assessments, others may show various and sundry subjective assessments by interested parties and another may show cost. The decisionmaker can then draw conclusions based on an integrated variety of information.

Scorecard analysis was originally applied to U.S. Department of Transportation policy in 1971 (Chesler and Goeller, 1973) and to environmental policy (Goeller et al., 1977). Davis has done extensive research in framing a problem for decisionmakers with scorecard methodologies (Davis, 2002; Davis, Shaver, and Beck, 2008; Davis, Johnson, et al., 2008). Scorecards are now ubiquitous in DoD and commercial practice.[6]

In a basic form of MCDA, the scorecard may be the final product of the analysis. The decisionmakers are then left with the task of assessing the extent to which their objectives are met by the entries in the matrix. The main feature of scorecard analysis we highlight is that the scorecard, or performance matrix, is directly analyzed by decisionmakers, i.e., the options remain separate and are intuitively compared and judged.[7]

Davis discusses what he calls static and dynamic scorecards. Static scorecards assign one weighting to each criteria and have a single product. Dynamic scorecards involve changing the relative weights among criteria to see how the scorecard rankings change. This essentially enables the decisionmakers to do a measure of tradespace exploration. Hillestad and Davis (1998) demonstrate a spreadsheet tool called DynaRank that allows a user to build such a dynamic scorecard.

Applying the Scorecard Approach

This approach could be applied in many ways. One way is to simply list the projects and their scores for each criterion. That would reflect essentially no mission orientation or connection (aside from what one could possibly infer from each project's location). Perhaps the way to reflect mission orientation with the least information processing or synthesis would be to add one criterion to the project list, the main mission the project supports. One way to display this could be to take the projects in the project tradespace, display them by mission area, and allow the AFCS personnel to establish a level of funding based on that information alone.[8] This could be plausible, if few enough projects are in the tradespace and the criteria are sufficient to allow intuitive processing of all the information while mitigating cognitive burden.

---

[6] For example, see Iseler, 2003; Defense Business Practice Implementation Board, 2002; and Abel, 2015.

[7] In analytically more sophisticated MCDA techniques, the information in the basic matrix can be converted into consistent numerical values. This presupposes that stakeholders are likely to be indifferent between equally scored alternatives, so that good performance on one criterion can in principle compensate for weaker performance on another.

In this case, one can combine the values of each criterion into a single value term using a utility function. (This is also sometimes referred to a utility model, value function, value model, or other similar terms, depending on the field of study.) The utility function reflects the criteria of interest to decisionmakers (i.e., objectives) with their respective weights, which arrives at an overall value for each option (in this case, infrastructure project). One type of utility function is called a linear additive model. The linear model shows how an option's values on the many criteria can be combined into one overall value. This is done by multiplying the value score on each criterion by the weight of that criterion and then adding all those weighted scores together. The Air Force's project priority model shown earlier is a linear additive model.

[8] We have seen this concept written about in internal documentation and have discussed it with Air Force personnel.

This scorecard could also be presented dynamically, such that decisionmakers have a few levers or criteria they can vary and then view the resulting projects. Given the dynamics of POM deliberations, this would probably require personnel from the AFIMSC or installations panel to do some advance analysis, presenting a menu of options (i.e., courses of action from which to decide).

One of the key tensions with this approach that we highlight (and revisit later) is, as Hillestad and Davis (1998) put it, between "what can be accomplished 'technically' and what must be assessed by the decisionmakers themselves." Determining that appropriate balance requires intimate knowledge of the available information on Air Force infrastructure and the AFCS decisionmaking process and environment.

Single-Year Versus Multiyear Perspectives

In Chapter Two, we discussed the inherently multiyear nature of infrastructure planning, investment, and maintenance. Yet project proposals that inform the IPL (and would thus be included in the scorecard) are only developed for the most urgent needs for allocation of funding in the year of execution. Part of the advantage of the project scorecard approach is that the projects are tangible and specific. Eventually, every building needs its roof replaced, and one can forecast with some reliability how often, for different types of roofs, and for buildings in different types of environments. (The Air Force has computer models that do this, which we discuss below.)

But in a POM deliberation, the replacement of some roof on some building being postponed is not the same as *the* roof on maintenance hangar #1 at Langley AFB for the 1st Fighter Wing (for example). Not all projects are so compelling as this example, but virtually any specific project is more compelling than a general description of a similar task.

POM deliberations, while certainly focusing on the first year of the FYDP being considered, in fact allocate money across the entire FYDP, and the effects of that multiyear allocation, especially for infrastructure, take time to manifest. Insufficient funding next year will not degrade condition all that much, but over several years can have a significant impact.

What this means is that the project scorecard approach may not be all that useful for communicating long-term impacts. In Chapter Four, we discuss how the weaknesses of each approach can be mitigated. We propose a way for the scorecard approach to be combined with others to capitalize on its potential potency while accommodating its near-term perspective.

Mission Outcome Metrics

The second approach is to quantify the impacts of asset performance on mission outcomes using targeted mission metrics. We describe three examples here of past and ongoing analyses.

Global Positioning System

Snyder et al. (2007) did a version of this mathematical modeling for the Global Positioning System (GPS). They explored the effect of reduced maintenance funding for GPS ground control stations on the accuracy of the GPS signal. They tied infrastructure asset performance to mission performance but stopped short of quantifying the full connection between infrastructure funding and mission performance. Figure 3.2 shows one of the outputs for their analysis.

In Figure 3.2, the key input is the mean time between critical failure (MTBCF) of the ground stations sending updates to the satellites, i.e., the performance of an individual infrastructure asset. The key mission performance metric is the 99th percentile (rather than the average) of mean estimated range deviation (ERD), one measure of the accuracy of the signal to a ground receiver. This was chosen to be especially sensitive to smaller changes in infrastructure performance.

As involved as this example was (requiring a simulation of how GPS signals decay and are restored as they receive updates), it was fairly simple in that only one type of infrastructure asset was used, mission performance could be distilled to a single metric, and data sources already existed that could be used to develop mathematical models. Many of the Air Force's other missions and bases are not so simple or straightforward.

Figure 3.2. Example of Global Positioning System

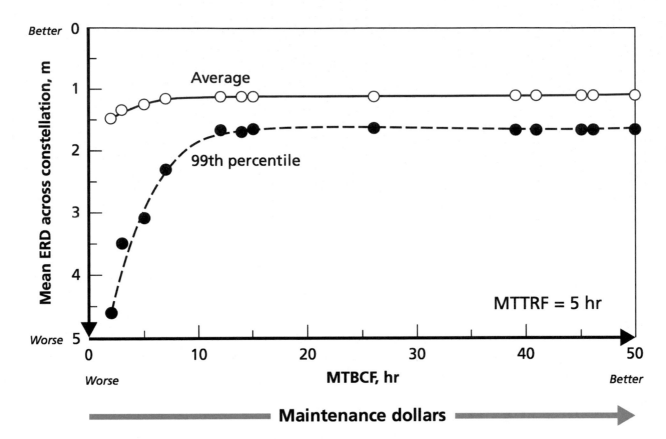

SOURCE: Snyder et al., 2007.
NOTES: MTTRF = mean time to repair functionality, m = meters.

Aircraft Sortie Generation Under Base Attack

Another example from recent RAND research is probably more analogous to most Air Force bases. Since FY 2012, PAF has been conducting research into resilient operational capabilities in denied environments. To support that effort, PAF has developed a suite of tools, one of which is called the Theater Airbase Vulnerability Assessment Model (TAB-VAM) (Thomas et al., 2015).

TAB-VAM models the effect of missile attacks on airbases. It incorporates the potential contribution of seven resource areas on the capability of the base to generate aircraft sorties (its mission metric). The resource areas are fuel storage, maintenance equipment, aircraft (vulnerable when parked in the open), runways, maintenance personnel, missile defenses, and munitions. Two of these are infrastructure assets or systems: fuel storage and runways. They model these with linear relationships between the availability or capacity between the infrastructure system and sortie generation.

Pilot Training

A somewhat different example is the Predictive Readiness Assessment System (PRAS), overseen by AF/A3O.[9] PRAS is not an infrastructure model; it focuses on pilot training.

PRAS is a predictive tool used to assess current readiness and the impact on readiness of alternative policies, force structure changes, contingency plans, funding cuts, and operational tempo. PRAS has been used to assist the air staff with a series of cuts and plus-ups in flying hours and to assess the potential impact on training of aircraft maintenance issues and of changing Air Expeditionary Force rotations from a 120-day to a 179-day rotation.

One relevant aspect of PRAS is that it uses metrics that are somewhat abstract, but still speaks the language of the core stakeholders. It categorizes outcomes the way the Status of Resources and Training System (SORTS) does: personnel, equipment, supply, and training. But it uses index metrics to synthesize what are inherently complex phenomena (e.g., the way a range of individual training tasks contribute to overall pilot readiness status). One can observe the vector of each metric (rate and direction) to understand whether it is getting better, getting worse, or staying more or less constant. One can quantify the time to get well after some deficit is incurred and understand the amount of money necessary to get there.

Also, PRAS integrates several input parameters—the Flying Hour Program, weapon system sustainment (i.e., depot maintenance), critical skills availability (i.e., field-level maintenance manpower mix), training resource availability (e.g., flying ranges), and deploy-to-dwell ratio (i.e. spin rate—and addresses the interactions among these variables.

One advantage PRAS has is that it has tangible inputs, such as flying hours, that easily translate to sorties, which translate to training tasks, which can be synthesized. So some (but not all) of the linkages are clear. PRAS is extremely detailed and sophisticated (and complicated) and took many years (and significant funding) to develop, so it serves as an extreme example of what is achievable for this kind of model.

Models developed to implement this method could be designed around a single asset, a group of like assets (in the case of the GPS above), or an entire infrastructure system, depending on how many assets have projects up for deliberation (or, in the case of multiple assets, on a single project) and what makes sense as a modeling approach (e.g., it would not have made much sense to model a single GPS ground station because of the dependency of signal accuracy on the other stations across the globe).

Single-Year Versus Multiyear Perspectives

The mission outcome metrics approach using computer models is neither inherently single- nor multiyear in its time horizon. The GPS example described above was run as a steady-state system, i.e., each funding level results in a constant level of performance from the infrastructure

---

[9] PRAS was developed by Alion Science and Technology. PRAS is being expanded to include space and unmanned aerial system assets, as well as automation of community-wide assessments.

assets in question (the ground stations). That single level of infrastructure performance results in a single (though statistically variable) GPS signal accuracy value. In reality, an insufficient level of infrastructure maintenance funding would lead to a gradual (though perhaps not elegant) degradation of the performance of the ground stations. That would then translate to a gradual degradation in the signal accuracy over a potentially long time horizon. Any similar model that predicts operational performance based on infrastructure performance can be used to generate multiyear effects, assuming that the infrastructure performance degradation can be reasonably predicted or modeled.

Composite Risk Metrics

The third approach we explore is to present composite risk metrics in the AFCS. This entails translating asset-level infrastructure performance into some high-level risk framework. For example, the Joint Chiefs of Staff (JCS) risk matrix has two dimensions: the type of risk (risk to mission and risk to force) and level of risk (low, moderate, significant, and high).[10] It is beyond the scope of this report to weigh the strengths and weaknesses of different risk frameworks. Here we describe a process by which the Air Force can arrive at such composite risk metrics.

To do so, we describe a four-step process.

- **Develop asset performance metrics.** How well is an asset performing its current function/mission?
- **Determine performance thresholds.** Assign thresholds (or ranges) that indicate when performance is not meeting mission.
- **Determine mission criticality.** The risk implications of various levels of performance may be different for different missions.
- **Translate thresholds and criticality to a risk framework.** To arrive at the proper risk framework, one must define that first, as the rest of the data gathered and analyzed have to support that choice. But we explain them in the order above.

Infrastructure Asset Performance Metrics

Infrastructure performance refers to how effectively, safely, and efficiently an infrastructure asset performs its mission at any time during its life cycle. This performance state, which changes during time in service, is reflected by two different indicators: the physical condition state and the functionality state. The physical condition state relates to a facility's general "physical fitness," and the functionality state relates to the facility's suitability to function as intended and required for the mission.[11]

---

[10] As described in Gallagher et al., 2016.

[11] Grussing et al., 2010a. The original definition, drawn from the American Society for Testing and Materials (ASTM) E 1480 Standard—Standard Terminology of Facility Management (Building-Related), was specific to buildings only. We adapt this definition to apply broadly to all types of built infrastructure.

In practice, the Air Force, Army, and Navy all define asset performance similarly, as does the National Aeronautics and Space Administration (NASA). In this section, we discuss the infrastructure performance concepts and metrics for these four organizations to set up a discussion about assigning performance thresholds. Figure 3.3 shows the basic terms in the performance definitions for each organization.

Figure 3.3. Service and NASA Models of Infrastructure Asset Performance

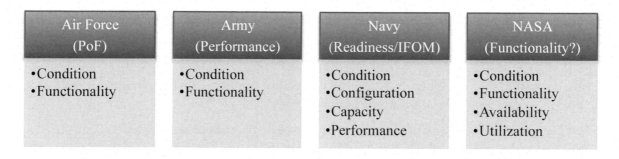

SOURCES: U.S. Air Force, 2015; Grussing et al., 2010a; Streicher, 2008; Dunn and Sawyer, 2013.
NOTE: IFOM = installation figure of merit.

All four organizations use similar definitions of *condition*, utilizing CIs similar (sometimes identical) to those described in Chapter Two in the discussion of the Air Force's prioritization model. CIs vary across types of infrastructure depending on their attributes, and also within types of infrastructure, given different levels of detail and rigor. Irrespective of an infrastructure asset's purpose, the condition of infrastructure assets (and their subcomponents) can be quantified and articulated in ways that incorporate the collective expertise of engineers.

Definitions of *functionality* vary more than those of *condition*, both within and outside our four example organizations. The Army uses the Facility Functionality Index (FFI), developed by the U.S. Army Engineer Research and Development Center (ERDC) and since incorporated into BUILDER (Grussing et al., 2010b).

According to ERDC, building functionality loss is a result of one of three factors (Grussing et al., 2010b):

- *User requirements:* change in tenant requirements or the underlying designated mission
- *Codes and regulation:* new building codes, regulations, or organizational policies
- *Materials and technology*: as a result of improvements to efficiency, maintainability, and overall performance of buildings, existing building components become obsolete and have lower capabilities in relation to the new baseline.

The Army's Functionality Index (FI) framework has two dimensions: functionality categories and criticality. Functionality is assessed along 65 specific functionality issues, which are grouped into 13 functionality categories.[12] Criticality is assessed along two dimensions: Severity quantifies the effect, and density quantifies how widespread the issue is. This is a data-intensive approach, but is quite comprehensive and rigorous.

The Air Force also uses BUILDER, and so has the same functionality index available (and includes the elements of the FI in its prioritization model definition), but at the time of this research had not yet fully implemented its features. We discuss this more below.

In the Air Force, condition and functionality are rolled up into one CI for the prioritization model. The Air Force allows further inclusion of the concept of facility function with a matrix consisting of two dimensions: A set of ten "functional purpose categories"[13] helps determine the functional purpose of a facility, and a functionality score assigns one of four levels of mission criticality to the loss of functionality.[14] Project proposers must provide functionality information and obtain permission from the local facilities board to use functional changes to inform the PoF term in the prioritization model (U.S. Air Force, 2015).

The Navy assesses infrastructure asset performance using a readiness metric called installation figure of merit (IFOM), with four components. Condition is the first, mentioned above. Configuration is essentially functionality and is calculated based on the proportion of space coded according to three levels of performance: adequate, substandard, and inadequate. Capacity measures the percentage of existing facilities that meet basic facility requirement authorizations within a facility category code at a site. And performance, measured by "capability performance level," quantifies the annual operating performance level as reported by installation commanders through a Navy quarterly reporting system (Streicher, 2008).

One proposed approach for NASA shares similar features (Dunn and Sawyer, 2013). In this definition, functionality is the ability of a facility to meet its intended purpose in terms of mission support, according to eight categories, analogous to the Army's FFI.[15] Availability is the readiness of the facility to provide intended services at any given time, whether it is actually occurring or not. It is a function both of system reliability and designed capacity. Designed capacity is the current planned availability; capacity can be constrained below original facility design levels by constraining resources required to operate (such as staffing or electrical power).

---

[12] Documentation shows there are now 13 major categories: location, building size and configuration, structural adequacy, access, Americans with Disabilities Act compliance, antiterrorism and force protection, building services, comfort, efficiency and obsolescence, environmental and life safety, missing and improper components, aesthetics, and maintainability. An early version included cultural resources as a 14th category (Grussing et al., 2010b).

[13] Air Force Instruction 32-1032, 2014. The ten categories are administrative, industrial, housing, medical, storage, education and training, community support and recreation, airfield pavement, grounds, and utilities.

[14] The functionality score follows the typical 100-point scale: 100 points when absolutely no workarounds are available and other options are more costly than the proposed action; 80 points when a workaround is available but the mission will remain degraded; 60 points when a workaround is available and the mission will not be impacted (but not all policies may be met); and 40 points when it is merely a quality of life issue but has no mission impact (U.S. Air Force, 2015).

[15] These eight categories are safety, legal (code compliance), environmental, energy efficiency, asset stewardship, staff readiness, product quality, and operation quality (Dunn and Sawyer, 2013).

Reliability, largely a function of condition, captures disruption to operations. NASA also captures utilization in its performance assessments but does not use it as a discriminator for project prioritization.

The key in all this is that these organizations all share strongly overlapping definitions of performance, mainly driven by condition and functionality.

Infrastructure Asset Performance Thresholds

To translate infrastructure asset performance to risk, performance thresholds must be assigned. Some of these thresholds already exist. For example, engineering-research-based CIs[16] measure the physical condition of facilities, their systems, and their components. These types of indexes are based on empirical engineering research and are the driving engines for SMSs like BUILDER.[17] Points are deducted from the 100-point index based on type and severity of component distresses. For example, a "high" severity generally denotes health, life-safety, or structural integrity problems or mission impairment. Inspectors merely collect distress data, and they do not make judgments concerning physical condition. The indexes have been applied to airfield pavements, roads and streets, railroad track, roofing, and building components (NRC, 2012).

Functionality thresholds, on the other hand, should be informed by users. Tools like BUILDER have automatic functionality triggers, but users (of different mission types, in particular) may have specialized needs, which can be specified in the SMS itself. Here, the Air Force's functionality lexicon is useful. It defines four levels of criticality that indicate how much a change in function affects the facilities' activities (U.S. Air Force, 2015):

- 100 points: Absolutely no workarounds are available, and other options are more costly than the proposed action.
- 80 points: Workaround available; mission will remain degraded.
- 60 points: Workaround available; mission not impacted. Not all policies met.
- 40 points: Nice to have, quality of life.

The Air Force uses the quantitative values to capture changes in functionality, but the language could be useful for articulating to a high-level audience the level of mission impact from a loss of functionality.

---

[16] This is as opposed to condition indexes based on maintenance backlogs, like the current index system used by DoD for real property inventory.

[17] Each index follows a mathematical weighted-deduct-density model in which a physical condition-related starting point of 100 points is established. Some number of points is then deducted on the basis of the presence of various distress types (such as broken, cracked, or otherwise damaged systems or components), their severity (effect), and their density (extent). The deductive values were based on a consensus of many building operators, engineers, and other subject-matter experts.

Infrastructure Asset and Mission Criticality

The models of performance assessment described above do not differentiate among the types of missions or their relative importance. Another tool that can be used to assign performance thresholds (including both condition and functionality) is mission criticality. Here we highlight two mission criticality metrics currently used in DoD for infrastructure.

Mission Dependency Index

MDI was developed jointly by the U.S. Navy, the U.S. Coast Guard, and NASA as a process for incorporating operational risk management into facilities asset management. As originally developed, MDI considers the ability to withstand mission interruption and the ability to relocate a mission to another facility. That is, if a facility or component is deemed not usable for mission accomplishment, for how long will the mission be interrupted (minutes or days?), and can the mission be moved elsewhere (is it impossible or easy)? The Navy[18] and Army (Grussing et al., 2010a) use extremely detailed MDI assessments, populated by structured interviews with stakeholders, including weights assigned to the various elements of the metric (NRC, 2012).

The Air Force attempted to apply this methodology at Langley AFB but found it to be too cumbersome and costly to implement service-wide.[19] As a result of Air Force interviews at Langley, the initial implementation team utilized a simplified approach: They took the MDIs elicited at Langley for specific facilities, assessed their DoD CATCODEs, and then applied those CATCODE-to-MDI rules more generally. This was intended to be a stopgap solution, with later efforts intended to refine those numbers.[20]

Because of this history, MDI in the Air Force has not always been an accurate measure of an asset's importance to the mission. Guidance now encourages base-level programmers to address MDI discrepancies with AFCEC if they see that there is an error in the assigned MDI (U.S. Air Force, 2015). If adequate justification is provided, the MDI will be changed to a more appropriate value.

For reference, Table 3.2 shows the Air Force's MDI levels.

---

[18] In its original form, a local installation commander and staff, via structured interviews, collect information from stakeholders on mission relocation capability and mission tolerance for interruption, considering both facility intra-dependency (within a mission) and interdependency (between missions). For more information, see U.S. Navy, Naval Facilities Engineering Service Center, n.d.

[19] For reference, U.S. Army ERDL estimated that "a full MDI assessment costs an average of $1,500 per mission sub-element on the installation. Typically, installations have 25–50 mission sub-elements. Therefore, the cost of a traditional MDI assessment is estimated at $40K–$75K per installation" (Grussing et al., 2010a, p. v).

[20] Email conversation with personnel from AFIMSC/IZBS on March 18, 2016.

Table 3.2. Air Force MDI Levels

Tier	Criteria	Examples
1	Mission critical, roughly top 10%, with recommended MDI from 85 to 99	• Example 99: operational runway, space operations facility, jet fuel storage • Example 85: deployment processing facility, air passenger terminal
2	Direct mission support, roughly top 25%, with recommended MDI from 70 to 84	• Example 84: R&D laboratories, primary water/wastewater assets • Example 70: maintenance hangar, technical training classroom
3	Base support, roughly top 50%, with recommended MDI from 60 to 69	• Example 60: test stand, education center, propulsion engine test cell
4	Community support (no mission impact), with recommended MDI below 60	

SOURCE: U.S. Air Force, 2015.
NOTE: R&D = research and development.

Critical Infrastructure Program

The Air Force Critical Asset Risk Management (CARM) Program[21] is part of a suite of programs and disciplines within the Air Force that address contingency planning, risk management, and mission assurance plans, such as emergency management, anti-terrorism, continuity of operations, and readiness plans. The CARM program establishes a comprehensive Critical Infrastructure Program (CIP) to identify, assess asset criticality, prioritize, and protect critical Air Force cyber and physical infrastructures.[22]

Under the CARM program, threats, hazards, vulnerabilities, and risks to Air Force–owned critical infrastructure needed to support mission requirements are captured in Air Force Critical Asset Risk Assessments (CARAs).[23]

The CARM program breaks criticality of infrastructure into four tiers:[24]

- Tier I—Warfighter/combatant commands suffer strategic mission failure. Specific time frames and scenarios assist in infrastructure prioritization.

---

[21] Air Force Policy Memorandum (AFPM) to Air Force Policy Document (AFPD) 10-24, 2012. AFPD 10-24, *Air Force Critical Infrastructure Program (CIP)*, implements Department of Defense Directive 3020.40, *Defense Critical Infrastructure Program (DCIP)*, 2010, and Department of Defense Instruction 3020.45, *Defense Critical Infrastructure Program (DCIP) Management,* 2008. The AFPM issued in 2012 renamed the CIP to the Air Force Critical Asset Risk Management (CARM) Program to reflect the risk management focus of the program on the relationship of critical infrastructure to the mission. The policy memorandum includes a statement indicating that the policy "becomes void after 180 days have elapsed from the date of this memorandum, or upon incorporation by interim change to, or rewrite of AFPD 10-24, whichever is earlier." It does not appear that an interim change or rewrite of AFPD 10-24 was issued within this time frame, which suggests that AFPD 10-24, issued in 2006, is the current policy and that the program name change from CIP to CARM no longer applies.

[22] AFPM to AFPD 10-24, 2012.

[23] As mandated by DoD and Section 335 to the National Defense Authorization Act of 2009 (Public Law 110-417).

[24] AFPM to AFPD 10-24, 2012.

- Tier II—The Air Force suffers mission failure, but warfighter strategic mission is accomplished.
- Tier III—Individual element failures, but no debilitating strategic or Air Force mission failure.
- Tier IV—Everything else.

The CARM tiers could be used in combination with, or separate from, MDI to assign levels of risk to degraded infrastructure performance.

Apply Information to Risk Framework

The final step is to translate infrastructure assets of varying criticality at various performance thresholds into some kind of distilled risk framework. One attractive possibility is a framework developed by AF/A9. The AF/A9 risk framework is based on the JCS's, for broad application to the Air Force (Gallagher et al., 2016). Like the JCS framework, AF/A9 includes risk to mission and risk to force; AF/A9 also suggests adding a category for risk to institutional objectives. AF/A9 includes four levels from the JCS: low, moderate, significant, and high.[25] AF/A9 then assigns percentage values to the levels of risk, in accordance with the JCS framework. Figure 3.4 shows this framework.

Figure 3.4. Depiction of AF/A9 Risk Framework with Common Air Force Thresholds

Risk Metrics	LOW	MODERATE	SIGNIFICANT	HIGH
Quantitative (Performance, Resources, or Schedule)	Metric value within 20% range of defined success	Metric value in 20% to 50% range of defined success	Metric value in 20% to 50% range of defined failure	Metric value within 20% range of defined failure
Quantitative	80% probability of achieving objective	50% to 80% probability of achieving objective	20% to 50% probability of achieving objective	0% to 20% probability of achieving objective
Range is the distance from defined success to defined failure points				

SOURCE: Gallagher et al., 2016.

Figure 3.4 shows the quantitative values that have been assigned to each level of risk in general practice in the Air Force. The top row includes a direct quantitative assessment of one or more metrics, e.g., performance, resources, or schedule. Performance could be something like aircraft availability, infrastructure condition, or on-time delivery of spare parts. Resources could simply be the percentage of stated requirements that are provided, e.g., number of fuel trucks for a fuel flight on a base. And schedule could be assessed in terms of lateness: A 10-month

---

[25] AF/A9 also adds upper and lower bounds, i.e., expected success and planned failure.

production process that was one month late would be low risk, 2–5 months late would be moderate, and so on.

The bottom row is also quantitative, but, instead of direct assessments, incorporates subjective or objective probabilities of achieving an objective. These could be produced by a model, elicited from subject-matter experts, or derived from historical data.

The key question of how, exactly, to translate infrastructure performance thresholds into this risk language is beyond the scope of this report. One possibility is to use the criticality rankings to designate the type of risk—mission, force, or institution—and assign levels of risk based on asset performance metrics (e.g., in the way Figure 3.4 shows metric values). Alternatively, the levels of risk could utilize the criticality ratings, such that degradation in a higher criticality asset would score higher in the level of risk.[26]

We do not attempt to lay out a comprehensive methodology to assign these risk ratings. We believe that such a translation requires input from engineers, infrastructure users, and decisionmakers to arrive at a lexicon that meets the Air Force's needs and is comprehensible and intuitive enough to all. The first question should be whether the outputs of such an effort would be beneficial in the POM deliberations. If the answer were yes, further analysis could be done to developing the performance to risk translations.

Populating Data

Some of the data and data capabilities necessary to implement this approach already reside in SMSs used by the Air Force. BUILDER, for example, has two capabilities that are not currently in active use that could satisfy some of the data needs for this analytic approach. BUILDER has the FI described above and a "Performance Index" that, as of yet, are not in use in the Air Force. The Air Force is working toward developing standard thresholds for the use of both the Facility and Performance Indexes in the future. Implementing these data capabilities requires significant manpower and will take time.

Single-Year Versus Multiyear Perspectives

The composite risk metrics approach is also not inherently single- or multiyear in perspective. SMSs like BUILDER and PAVER can generate multiyear predictions of facility condition under various funding scenarios. For instance, the Integrated Multiyear Prioritization and Analysis Tool (IMPACT) within BUILDER^{TM} can calculate the different amounts, types, and costs of deferred maintenance work that accumulate for a given asset over time if that asset is funded at different levels.

AFCEC provided a sample BUILDER^{TM} data set to the research team for Columbus AFB facilities. BUILDER calculates a condition index for each asset (58 buildings with dozens of

---

[26] We were told that the Air Force is currently working to couple condition and MDI tier to articulate risk to guide funding. (Feedback provided by personnel from AFCEC-CP, November 17, 2015.)

subsystems and thousands of components in the data set). The tool starts with an initial CI based on the asset's age and/or user input based on physical inspection. Each asset has a list of scheduled maintenance tasks to be performed over time, each requiring an amount of funding. With full funding, all the tasks are accomplished, and no deferred work backlog builds up for that component, system, or building. With less-than-sufficient funding, deferred work builds up, and the CI degrades quickly over time.

In the model, the component CIs are rolled up to the building level, and each building is given a single CI. Each building has an associated MDI (based on DoD Real Property Categorization System [RPCS] codes), and those MDIs are stratified into five levels to assess mission criticality. Here are the five MDI levels, with example buildings:

- Low: bowling facility
- Relevant: chapel
- Moderate: mission support group complex
- Significant: flight simulator building
- Critical: control tower.

AFCEC personnel ran two example funding scenarios for us to produce sample data sets: one with no funding and one with annual funding at 1.5 percent of total component replacement value (which corresponds to a 67-year recapitalization rate).

Figure 3.5 shows an example from the data. This figure shows component-level outputs for the several components on the fire station: overhead rollup doors, windows, and lockers. For each component, the dotted line shows the CI degradation with no funding, a more or less smooth degradation as no scheduled work is done and a backlog accumulates. The solid line for each component shows the 1.5 percent funding scenario. One can see the same gradual degradation over time until the year 2016, when all components experienced some kind of scheduled maintenance activity, raising their CIs significantly. Then they again gradually degrade over the rest of the time horizon. Not all components and buildings experienced the same jumps, and at a base level, both the zero and 1.5 percent cases roll up to a smooth CI decay over the ten-year period.

Figure 3.5. Example BUILDER-Generated Condition Index Data for Columbus AFB Fire Station Components

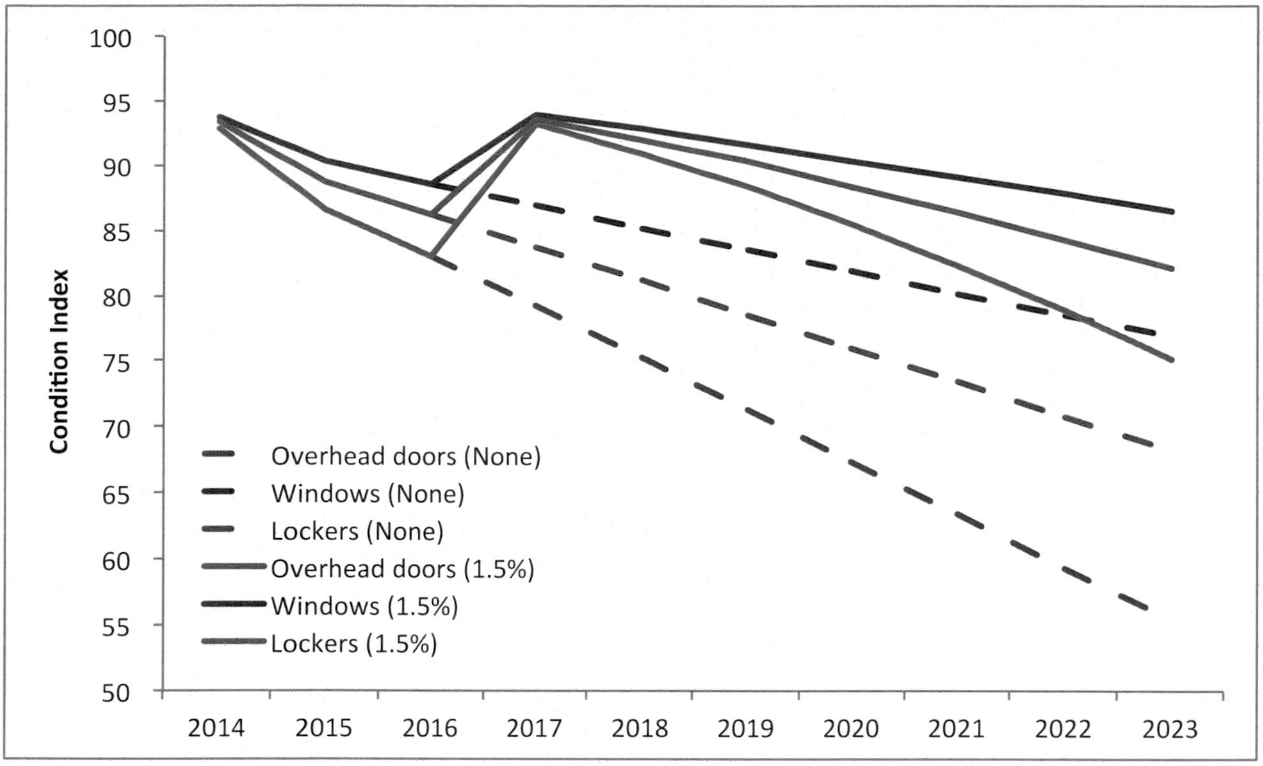

SOURCE: BUILDER, provided by AFCEC.

This kind of display (at a higher level of aggregation, of course) could powerfully communicate the long-term impacts of infrastructure underfunding. Were the Air Force to invest the time and money to fully populate condition and performance data in the systems and assign mission-based performance thresholds, generating displays such as these would simply be a matter of data analysis.

The Navy's model for allocating infrastructure funding, the Shore Facility Infrastructure Model (SFIM), has multiyear displays. SFIM, given a funding level over the FYDP, displays outcomes for the four performance metrics described in Figure 3.3. SFIM incorporates all assets and all types of investment. It also has a long-range forecasting tool that does the same across a 20-year time horizon (i.e., the value of each metric a the end of the time horizon), though it works at a lower level of fidelity than calculations for the FYDP, given greater uncertainties over such a long time.

A Case Study for This Project

We understand the data requirements and availability fairly well for the scorecard and composite risk approaches. For the mission outcome metrics approach, we performed a case study, presented in the next chapter, to explore the current availability of data to populate our framework and, by extension, the feasibility of applying our framework broadly to answer the research question. We chose two bases: Columbus AFB, within Air Education and Training Command (AETC), and Schriever AFB, within Air Force Space Command. We chose Columbus AFB for two reasons. First, AFCEC had already begun populating data in BUILDER for Columbus AFB, so we could use it for realistic cost calculations.[27] Second, because it is a small base with only one mission and few tenant units, we expected it to be straightforward.

We chose Schriever AFB because personnel there had begun adjusting the MDI values for their infrastructure assets to better reflect local knowledge (we discuss MDI more in Chapter Four), and because Schriever AFB does not have a flying mission, so we expected it to present challenges that Columbus AFB would not.

We only did a detailed case study of Columbus AFB. We also did a site visit to Schriever AFB, but the intent was mainly to see what was different and to identify and mitigate any obstacles. Chapter Four summarizes the results of the Columbus AFB case study.

---

[27] BUILDER is one of several management tools (provided by SMS) used by the Air Force to track its infrastructure inventory and condition. BUILDER is used for buildings, PAVER for pavements, and ROOFER for roofs.

4. Linking Infrastructure to Missions with Mathematical Modeling

In this chapter, we describe our analytic approach to the building models to quantify mission outcomes of infrastructure funding and apply our approach to a case study to illustrate it.

Analytic Approach

Our analytic approach has two steps. First, we built a logic model of a mission at a base. (A logic model is usually a graphical depiction of the logical relationships between the resources, activities, outputs, and outcomes of a program [Alter and Murty, 1997].) This requires answering three questions.

The first question is *What?* What are the elements of infrastructure and of readiness (i.e., mission performance) that must be linked? Populating this element of the framework entails cataloging infrastructure assets and articulating mission performance metrics.

For this task, we sought data on infrastructure assets and found plentiful data organized by infrastructure categories (CATCODEs), by which engineers manage assets (bottom of Figure 4.1). We also sought mission-oriented metrics against which to measure performance. The primary mission of Columbus AFB is to train pilots. Mission performance can be defined as the number of pilots that it graduates each year, i.e., pilot production rate (top of Figure 4.1). We found, however, that available data do not link the infrastructure assets and mission metrics in a way that enables quantification of the relationships (though local BCEs and operators usually know the linkages intuitively), leading to our next step.

The second question to answer in constructing the logic model is *How?* How do these elements relate to one another? This entails linking infrastructure assets with mission performance through some kind of structure and describing their connections—in other words, building a logic model. We introduce two new concepts to do this:

- Mission function: an activity or output that, together with other mission functions, produce the ultimate mission performance outcomes
- Infrastructure systems: groups of infrastructure assets that are grouped according to function (e.g., runway system) that can be mapped to mission functions.

For this task, we sought two types of data and information. First, we sought to decompose missions into their component mission functions, where applicable. For an Air Force serviceman to graduate as a pilot, he or she must have completed a specific number of flight hours, academic hours, and simulator hours. Thus, in the case of Columbus AFB, the mission functions we proposed are the academic function (traditional classrooms), simulator function (flight simulators and their facilities), and flight function (including runways, aircraft, fuel, etc.). Each element can operate essentially independently (i.e., a complete outage in one would not materially affect another), but without all three, Columbus AFB cannot fulfill its mission and graduate a class of pilots.

Third, we sought to answer the question *How much?* How much do changes in resource inputs (infrastructure funding) impact readiness outputs (mission performance metrics)? This entails building *mathematical* models to accompany the *logic* models.

Figure 4.1 shows these elements schematically. Starting from the bottom, infrastructure assets can be grouped into infrastructure systems, which can be mapped to mission functions, which collectively produce mission performance. The "what" of our framework is the top and bottom boxes of this diagram, the mission metrics (outputs) and infrastructure assets (inputs). The "how" element is the linkages between these elements, and the "how much" element is the mathematical relationships underlying those linkages.

Figure 4.1. Analytic Framework Linking Infrastructure Resources to Readiness

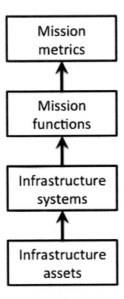

In the rest of this chapter, we describe in detail the Columbus AFB case study. As we walk through the elements of the framework, we further break them down into constituent tasks and identify the information required to complete these tasks. Some of the required information is already available within the Air Force, in which case, we use what is available in the case study. For necessary information that is currently not available within the Air Force, we point to potential sources to complete the case study.

Case Study for Columbus AFB

Metrics for Mission Performance

The primary mission of Columbus AFB is to train pilots. Mission performance can be defined as the number of pilots that it graduates each year, i.e., pilot production rate. For an Air Force officer to graduate as a pilot, he or she must have completed a specific number of flight hours, academic hours, and simulator hours. Hence, mission performance is a function of the availability of flight hours, academic hours, and simulator hours (see Figure 4.2). Each mission contributor is, in turn, a function of the availability and utilization of various infrastructure assets, such as the runway, taxiways, control towers, simulator buildings, classrooms, etc. The capacity of Columbus AFB to generate these hours will determine the number of pilots produced.

Figure 4.2. AETC Mission Task, Supporting Mission Requirements, and Relevant Mission Performance Metrics

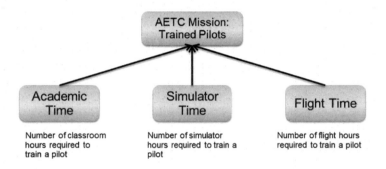

We used the training syllabi for T-6A Joint Primary Pilot Training, T-1A Joint Specialized Undergraduate Pilot Training (SUPT T-1A), and T-38C SUPT as references detailing how many hours students need in an academic environment, in a flight simulator, and piloting an aircraft before they could graduate from the listed programs. For example, to complete SUPT T-1A training, a student needs to complete 142.4 academic hours, 53.6 simulator hours, 11.6 ground training hours, and 76.4 flight hours. We then used historical production numbers to determine that, typically, eight cohorts are going through each of these programs in Columbus AFB at any point in time. Class size has been, and we can assume will be, limited to 30 for the T-6A and 21 for the T-1A. This tells us, for example, that having access to a classroom or academic system (classroom and associated infrastructure) for 142.4 hours can yield 21 SUPT T-1A graduates. Having such a system for less than 142.4 hours cannot yield any SUPT T-1A graduates.

Figure 4.3 presents an example of the mission-generation function at Columbus AFB for pilot training on the T-1 aircraft. The y-axis denotes our ultimate mission metric—pilot production. The x-axis shows the number of available hours for each of the three mission functions from Figure 4.2. Thus, for a given level of desired pilot production, one can derive the needed infrastructure asset performance (in available hours per time period) required to achieve it.

Note the step function linked to syllabus requirements; every time we have another 142.4 academic hours, we can (potentially) graduate another 21 students. We extended the analysis by considering attrition rate, the number of aircraft and simulators of various types present at Columbus AFB, the number of students that can simultaneously use aircraft and simulators of various types, etc.

Figure 4.3. T-1 Pilot Production Rate for Columbus AFB

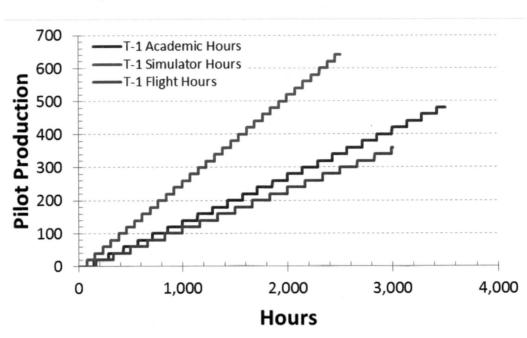

Of course, generating flight, academic, and simulator hours requires infrastructure systems and their assets: aircraft, fuel, runway, control tower, classrooms, simulator buildings, instructors, etc. The availability and performance of these infrastructure assets will determine the degree to which each mission contributor will be available to generate trained pilots.

Mapping Infrastructure Assets to Systems

We group the infrastructure assets at Columbus AFB into seven infrastructure systems: airfield, C2, base support, fuel, training support, aircraft support, and personnel support. We built this representation of the assets at Columbus AFB by manually mapping each asset in the RPCS database to each system. Figure 4.4 presents this mapping for the airfield infrastructure system and its constituent assets.[1]

---

[1] It was brought to our attention that airfield drainage and grounds maintenance were not shown in Figure 4.4. Both drainage and grounds maintenance are critical to delivering the end states described, but were not included in the real-property inventory data provided to us for Columbus AFB.

Figure 4.4. Mapping of Select Assets to Airfield Systems at Columbus AFB

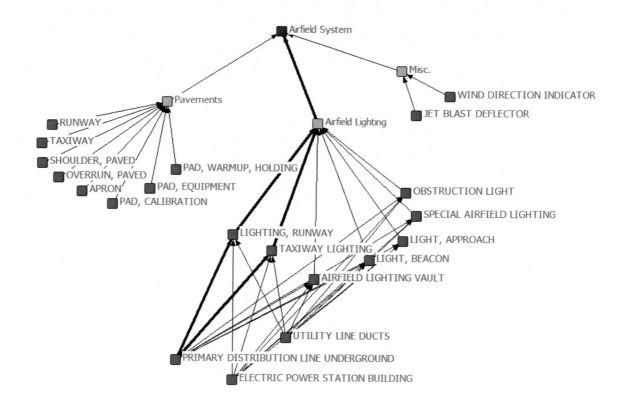

Note that, for ease of visualization and to facilitate the mapping exercise, we have also identified three airfield subsystems. Here they are pavements, airfield lighting, and miscellaneous. The pavements subsystem represents all airfield-related pavements, from the runways to taxiways, to aprons, to calibration pads; the airfield lighting subsystem represents all airfield-related assets that provide various types of lighting to the airfield; finally, the wind direction indicator and jet blast deflector are classified as miscellaneous subsystems that are part of the airfield infrastructure system. All lighting-related assets are also linked to electricity-related assets, which means that the operation of the airfield system also depends on such utilities-related assets as the power station and distribution lines. Table 4.1 provides the list of systems and their associated subsystems, along with an identification number that we use to show the mapping of the two-digit DoD asset codes for Columbus AFB to the subsystems.

Table 4.1. System and Subsystem IDs

System	System ID	Subsystem	Subsystem ID
Airfield Infrastructure System	1	Pavements	101
		Airfield Lighting	102
		Misc. Runway System Assets	103
C2 Infrastructure System	2	C2 Buildings	201
		C2 Radar and Navigation	202
Weapon Support System	3	Maintenance/Repair Buildings	301
		Storage	302
		Equipment	303
Fuel Infrastructure System	4	Fuel Storage	401
		Fuel Dispensing	402
		Fuel Misc. Facilities	403
Personnel Support System	5	Medical	501
		Housing	502
		Fitness	503
		Shopping	504
		Family Support	505
		Pet Services	506
		Recreation	507
		Dining	508
		Transportation	509
		Religious Services	510
Base Support System	6	Misc. Base Support	601
		Base Safety and Security	602
Training Infrastructure System	7	Simulator	701
		Training Aids	702

Based on the mappings we developed for Columbus AFB and Schriever AFBs, certain clear relationships appear between the assets and the subsystems. For example, most of the assets under the two-digit DoD Code 11 (Airfield Pavements) map to the Pavement Subsystem under the Airfield System (101). While one might expect all assets under the Airfield Pavements code to map to the Pavement Subsystem, there are exceptions; for instance, Aircraft Washrack (DoD Code 116672) maps to the Equipment Subsystem under the Aircraft System (303). Moreover, many single assets map to multiple systems and subsystems. For example, Water Distribution Mains (DoD Code 842245), which falls under the two-digit DoD Code 84 (Water), maps to numerous infrastructure systems: C2 Buildings (201), Medical (501), Dining (508), etc. The assets that supply water to an Air Force base should touch almost all of the support systems. Without such assets functioning properly, those subsystems would suffer from loss of functionality. Through this analysis, we found that the two-digit DoD asset codes typically map to multiple subsystems and systems. Table 4.2 provides the mapping by two-digit DoD and subsystem codes for Columbus AFB.

Table 4.2. Mapping of Two-Digit DoD Asset Code to Subsystems

Two-Digit DoD Code	Subsystem ID
11	101, 103, 303
12	401, 402, 403, 509
13	101, 102, 103, 201, 202
14	103, 201, 202, 507, 601, 602
15	601
17	201, 503, 602, 701, 702
21	301, 302, 303, 402, 403, 509, 601
41	401
42	302
44	602, 601, 504, 401
45	601
51	501
54	501
61	601
69	506, 601
72	502, 601
73	507, 509, 510, 602
74	502, 503, 504, 505, 507, 508, 601
75	507, 509
76	507
81	102, 103, 201, 202, 301, 302, 303, 402, 403, 501, 502, 503, 504, 505, 506, 507, 508, 509, 510, 601, 602, 701
82	201, 301, 402, 403, 501, 502, 503, 504, 505, 506, 507, 508, 509, 510, 601, 602, 701
83	201, 301, 402, 403, 501, 502, 503, 504, 505, 506, 507, 508, 509, 510, 601, 602, 701
84	201, 301, 402, 403, 501, 502, 503, 504, 505, 506, 507, 508, 509, 510, 601, 602, 701
85	302, 402, 509
87	509, 601, 602
89	102, 103, 201, 202, 301, 302, 303, 402, 403, 501, 502, 503, 504, 505, 506, 507, 508, 509, 510, 601, 602, 701

An example of an asset-to-asset mapping is the asset Primary Distribution Line Underground (CATCODE 812225). It maps to Taxiway Lighting (CATCODE 136667) and Lighting, Runway (CATCODE 136664), which in turn both map to the Airfield Lighting Subsystem (102), which itself maps to the Airfield Infrastructure System. The Primary Distribution Line Underground asset maps to many other assets requiring underground electrical power, such as the Store, Commissary (CATCODE 740266) asset, which maps to the Shopping Subsystem (504), which then maps to Personnel Support System and the Radar and Navigation Subsystem in the C2 System, among many others. As one would expect, assets that are essential to the mission and sustainment of the base (e.g., electrical power and water distribution) map to many other assets, which then in turn map to a large number of subsystems and systems.

The objective of this mapping was to demonstrate which systems that support Columbus AFB's mission are directly affected by the asset. While it is clear, for example, that the Airfield System could not function properly without the Road assets (CATCODE 851147) in good condition to support the day-to-day activity of the base (and, hence, they map to the Base

49

Support System), it does not directly affect the Airfield system in the same way a Taxiway asset (CATCODE 112211) does. This narrow construction of relationships between assets to systems avoids judgments about the necessity of assets for a system to be functional. Clearly, the Road assets would be necessary for almost any AFB system to be operational; conversely, the Rod and Gun Club (740315) is unlikely to be necessary for most systems to function. On the other hand, many other assets, such as the Youth Center (CATCODE 740883) or Dental Clinic (CATCODE 540243), might eventually affect other systems' capabilities; however, since they are not directly related, such mappings are not established in this analysis.

While the effort involved in completing such a mapping is nontrivial, it needs to be undertaken only once for a given base, then updated periodically as assets, units, or functions change. Further, we began from scratch, with little specific knowledge of the particular base. BCEs and operators, on the other hand, tend to have tacit knowledge of how assets fit together on their base, so the requisite level of effort on their part to complete this exercise is likely to be lower.[2]

Mapping Infrastructure Assets to Missions

Having identified the contributors to the mission and having grouped all the infrastructure assets into infrastructure systems, it is possible to create the final link between the assets and the mission (Figure 4.5). In this example, the AETC mission has three components: flight time, simulator time, and academic time. Training a pilot requires a specific number of each of these, as discussed in a previous section. To generate each of these time components, it is necessary that the seven infrastructure systems be available. Figure 4.5 suppresses the detailed asset to system mappings for legibility.

---

[2] We showed this network map to engineers and operators at Columbus AFB, and they agreed with both the approach taken to building this network map and with the actual groupings.

Figure 4.5. Assets to Mission Mapping for Columbus AFB

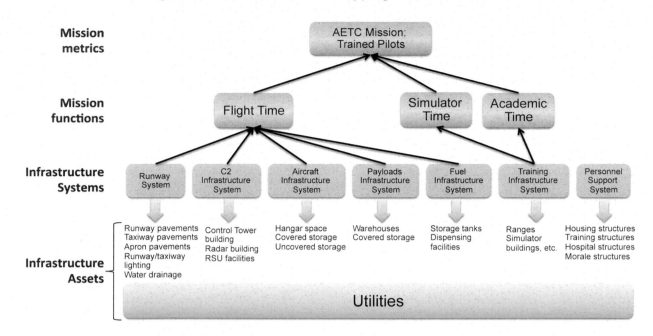

NOTE: RSU = runway support unit.

The airfield, C2, aircraft, payloads, and fuels infrastructure systems contribute to and enable the generation of flight time, while the training infrastructure system enables the operation of the simulators and classrooms that generate the simulation and academic time required to train pilots. For instance, the aircraft made available by the aircraft infrastructure system provide the aircraft that will fly the training sorties, the fuel made available by the fuels infrastructure system enable the operation of the aircraft, operation of the airfield system enables the takeoff and landing of aircraft, and operation of the C2 infrastructure system (i.e., control towers) enables the safe operation of the aircraft.

Similarly, operation of the simulator buildings and classrooms will determine the amount of simulator time and academic time that pilots can utilize, respectively. Note that we decided to not link the personnel support system to the mission because the assets that comprise this system are necessary for all the functionality of the other systems as well as the base itself; they play a similar role as the utilities, i.e., they enable the operation of the other systems. In implementing an approach like this, it is probably not cost-effective to map some of the infrastructure assets in what we termed the personnel support system (e.g., the youth center and dental clinic referred to above). While some of these may have a second- or third-order mission consequence, a modeling effort must choose elements to include that justify the time and trouble of doing so.

Besides identifying dependencies, one important element of the mapping exercise is to identify redundancies. Whether among infrastructure assets or entire infrastructure systems, redundant capabilities can mitigate risk.

The performance of each system and the service that it provides to the mission will be a function of the performance of the assets that constitute that system. This, in turn, will be a function of the funding available to repair and sustain each asset. The next step in the analysis is to quantify asset availability and its relationship to funding level.

Quantifying the Linkages—Developing Mathematical Models

Earlier, we described the task of quantifying the linkages between infrastructure funding and mission performance as two fundamentally different kinds of tasks: Linking funding to asset performance is a question of data and statistical inference; linking asset performance to mission performance is more a question of conceptual modeling (like the examples of other models given in Chapter Three).

Given the linkages provided in Figure 4.4 (and its more detailed version), conceptual and mathematical models could be created to tie assets to systems and systems to mission functions. For example, TAB-VAM already does this for some resources supporting flight time or sortie generation. Many of the relationships are simply linear, e.g., hours per day of operational runway time or capacity of the fuel system. Much of the modeling would probably have to be done at the infrastructure system or subsystem level (e.g., pavements and airfield lighting both supporting the airfield system), with some simple rules about how assets contribute to subsystems and systems.

Unfortunately, in our research into the availability of current data and models, we found the remaining task—modeling the effect of funding on individual asset performance—most problematic. The holy grail in quantifying readiness is to quantify actual mission outputs[3] or, in the case of an individual asset, that asset's outputs. But most research in this field is not oriented toward directly quantifying outputs. Among the major types of infrastructure the Air Force manages, pavement has the richest literature that could provide useful insights. But the study of facilities and utilities is much less mature, and there simply do not yet exist robust methods to estimate these relationships.[4]

The crux of the problem is that the literature to draw from is often insufficient. There is rich literature in some fields on condition (which we discuss in detail in Appendix A), but that does not always map to system availability or overall performance, and those are the elements most useful to inform models. The GPS and TAB-VAM models described in Chapter Three both use infrastructure asset availability and performance to inform system and mission outcomes.

This shortfall could be mitigated by focusing on the outcomes from a specific project. Engineers and users could identify the anticipated outcomes for an infrastructure asset or system that is the subject of a proposed project. For example, "If this project is deferred another year,

---

[3] See, for example, Harrison, 2014.

[4] Moreover, runways are the most easily funded, because of their obvious mission criticality. So there is probably the least risk of runways being underfunded and deteriorating, relative to, say, a gymnasium or a sewer line.

there is a low/medium/high likelihood that the asset/system will experience degraded performance, disruptions, etc." Then, those infrastructure outcomes could be modeled using the approach we outline in this chapter, quantifying the outcomes in terms of mission metrics.

5. Applying Methods to Air Force POM

In this chapter, we weigh the strengths and weaknesses of the three methods for linking infrastructure resources to mission for use in the POM discussed in Chapter Three.

Table 5.1 summarizes some strengths and weaknesses of the three approaches. In the rest of this chapter, we discuss those strengths and weaknesses, then describe ways to tailor each one to mitigate its weaknesses and make it more useful or viable in the POM process.

Table 5.1. Strengths and Weaknesses of Approaches

Approach	Strength	Weakness
Project scorecard	• Concrete, focus on single projects • Puts onus on mission owners • Little additional investment	• Could challenge bandwidth of AFCS • Could devolve to project focus • Could default to decisionmaking biases • Near-term perspective limits ability to express long-term implications of underfunding
Mission outcome metrics	• Concrete • Mission-specific • Output-oriented • Can identify unique input/output relationships • Compatible with displaying long-term implications of underfunding	• Potentially costly to implement • Each mission requires separate model • Still requires cross-mission assessment/trades • Not all missions may be amenable • Excludes more distant infrastructure activities like "municipality," personnel support
Composite risk metrics	• Risk framework already exists • Data systems support this approach • Compatible with displaying long-term implications of underfunding	• Metrics are more abstract • Requires investment of time and manpower to populate data (engineers to populate inputs; mission owners setting some thresholds)

SOURCE: Authors' analysis.

Project Scorecard

The scorecard approach presents some number of projects inside the tradespace (see Figure 2.1) that would be funded at some requested level, but would not be funded at a lower level. This focuses attention on what is lost if a lower level of funding is provided: All those projects will be delayed at least a year until the next funding cycle arrives. This display could involve all the projects in the tradespace or merely select projects that are illustrative and/or make the best case for more funding.

Advantages

One strength of this approach is that a single project is usually concrete: It is a singular physical task (or group of tasks) that can be envisioned. One can see and touch an infrastructure asset and project. People generally know how major infrastructure assets operate (a runway, a building, etc.) and can thus understand fairly concretely what a project would do.

Another strength is that, as envisioned, this approach probably involves mission owners (i.e., MAJCOM commanders or their representatives) advocating for projects that support their missions and justifying/explaining the mission impacts of delaying them. The mission owners know their missions and activities best and could (arguably) best articulate the effects. It is understandable that when presented as the installation panel's portfolio, and not as the global strike infrastructure portfolio, for example, the funding request would seem disconnected from Air Force missions.

Third, this approach is simple, in that it requires little additional data gathering or processing. It is likely that some preparatory analysis by the AFIMSC and possibly the installations panel would be necessary to pinpoint these projects and understand the effects, but probably only marginally more than is already done for the POM.

Disadvantages

There are several potential hazards involved in presenting individual projects. First, there could be an overwhelming amount of information and complexity in the project tradespace. If there were $1 billion on the table for all projects (a reasonable amount, given our analysis of funding data discussed in Chapter Two), every 10 percent of tradespace (say, between 80 and 90 percent of the request) would translate to $100 million. At the former minimum of $5 million per project, that should be no more than 20 projects within the tradespace. Given new thresholds, potentially hundreds of thousands of dollars, there could be 50, 100, or even more projects on the table for such a funding increment. Our notional project list in Figure 2.1 showed only about five or ten projects in each $100 million increment (in part for legibility); the actual list could have many more.

The next complexity could be the number of mission areas to assign to each project. Current guidance directs project proposers to assign one of three mission areas: Global Power, Global Vigilance, and Global Reach. Alternatively, there are five "core mission areas": air and space superiority; intelligence, surveillance, and reconnaissance; rapid global mobility; global strike; and C2. Or, there are 12 core functions, and scores of subfunctions (see Figure 5.1). Or, the ten MAJCOMs, as they are the mission owners. Having more mission categories makes the effect of a project more identifiable, but too many makes it challenging to intuit all the options.

The POM environment itself has dozens of formal members (plus stakeholders) ranging from operators and logisticians, to judge advocate general, the surgeon general, and chaplain. Having such a diverse group look at scores of projects from a dozen mission areas could easily be overwhelming.

A second disadvantage of this approach is that the mission impact of some projects may simply not be that compelling. Certainly, if the commander of Air Combat Command argues that reduced funding would delay the repair of a key runway for a fighter pilot training base, or a maintenance hangar needing renovation, the potential impacts are obvious, if sometimes uncertain. But the MDI summary in Table 3.2 suggests that even assets with moderately high MDI are not so attention-grabbing. The "mission critical" category (MDI 85–99) includes Operational Runway (MDI 99), Space Ops Facility (MDI 99), Jet Fuel Storage (MDI 99), Control Tower (MDI 90), Deployment Processing Facility (MDI 85), and Air Passenger Terminal (MDI 85). Even the "direct mission support" category (MDI 70–84) includes R&D Laboratories (84) and Technical Training Classroom (80). Many of the highest-MDI assets will not be in the tradespace up for discussion, because they will already be above the line. The tradespace could include many assets whose mission impact is simply hard to identify with. This could be mitigated by selecting high-impact projects from within the tradespace.

If the type and amount of information presented in the POM deliberations is, in fact, overwhelming, the deliberations could go in a number of directions. They could devolve into picking apart individual projects (Why can't this or that project wait another year?), scrutinizing the prioritization model itself (Why is that project not above the line?), or trying to make smaller trades among projects that seem intuitively appealing. Another possibility is simply that the deliberators could fall back on common decisionmaking biases and heuristics, as referenced in Chapter One.

A final disadvantage of this approach is that the near-term focus of the IPL (its purpose is to allocate a year of execution funding, not plan over the FYDP) limits the utility of the approach to express the long-term implications of infrastructure underfunding.

Mitigating the Weaknesses

It seems that the conditions that are best for this approach are when there are few enough projects, little enough complexity, and straightforward enough outcomes for decisionmakers to understand intuitively. If that is in fact the situation, some variant of the project scorecard approach could be cost-beneficial.

However, the more information and complexity there is, the more the presenters must somehow reduce it. One way is to simply tailor the project information highlighted for each funding bogey, i.e., the most compelling ones. If the line is drawn here, this gets delayed; if the line is draw further up, that gets funded. Decisionmakers might then ask, "So how much do we have to spend to get that project above the line?" This sort of discussion tends to end up with those responsible for mission areas like global strike doing a lot of the talking, and anyone in charge of areas like cyber being fairly silent.

One way to keep decisionmakers from picking apart the prioritization model is to build trust, so that stakeholders understand what goes into it (ideally having had input into the criteria and weights) and how it generates its results. Otherwise, each person can probably think of a way to "improve" the model. In other words, any funding level will probably leave behind a project that someone could argue should be above the line.

Another possibility is to combine the project scorecard approach with one of the other two approaches, to add risk/impact information to projects to make them more understandable. A variant of this would be to assign more risk or performance metrics to each project than the project prioritization process already provides. Risk could be communicated using some broadly applicable risk framework, with engineers and/or users providing risk information for each project.

Mission performance could be quantified and articulated in a number of ways. Base-level users or operators might already have the information and tools to quantify outcomes. For our case study of pilot training at Columbus AFB, the pieces of the puzzle were mostly provided by operations group personnel, who already used spreadsheets with mathematical formulas to track pilot production. We assembled that information into a single spreadsheet tool; users could use such an approach (whether crude or sophisticated) to quantify more clearly the mission performance degradations associated with delayed projects.

As for its near-term focus, this approach could be paired with one of the other two approaches. The project scorecard (and the ensuing discussion) could provide a potent way of articulating mission impact, albeit near-term, while another approach could provide a longer-term perspective. While project information could conceivably be developed for future years, this would probably only increase the likelihood of cognitive overload, something that already has to be managed with this approach.

Mission Outcome Metrics

This approach entails choosing useful mission metrics, then building logic and mathematical models to link and quantify the effects of infrastructure funding on these mission outcomes.

Strengths

Like the project scorecard, mission outcome metrics are, if well designed, concrete and relatable: sorties generated, pilots graduated, GPS signal accuracy, etc. They are also mission-specific, so they are very tangible. Finally, they are output-oriented, getting at what operators and decisionmakers are interested in and find compelling.

The model developed in Chapter Four is relatively crude, but even that allows someone to understand better how the subsystems contribute to Columbus AFB's pilot training objectives, and even to see that for a given number of pilots, the simulator system required significantly less operational uptime relative to the other two. Perhaps this means that the simulator system is more "interruptible" (in MDI language) and could withstand more disruptions than could the others. Building even crude models can reveal these relationships.

More sophisticated models (like the GPS model in Snyder et al., 2007) may also reveal interesting relationships, like a knee in the curve. In the GPS example, that was an explicit goal of the analysis—to find the proverbial cliff one falls off when funding is reduced too far. In addition, models like these can capture interactions among a number of variables that may be too difficult to intuit without such an aid.

This approach is also compatible with displaying long-term impacts of underfunding. Depending on how the model is structured, it might not even require additional calculations to show multiyear impacts. And a long-term perspective is key to showing decisionmakers the real dangers of underfunding.

Weaknesses

Unfortunately, these mission outcome models can be costly, in both time and manpower, to develop. For the case study in Chapter Four, much of the basic information and relationships had already been worked out by operations group personnel. It is their job to ensure that pilots get trained, so they had already worked out some of the math. The PAF research team gathered other data, then had to synthesize that into a spreadsheet model. A more sophisticated model would certainly be necessary to capture relationships with individual infrastructure systems, or even individual assets. This requires time and manpower, including analysts to build the models and engineers and operators to provide inputs.

Second, each mission potentially requires a separate model. One simply must look differently at flying aircraft missions, nuclear missiles, space-based missions, or even cyber activities. While many Air Force bases do fly training sorties as part of their mission set, each base may have important characteristics that need to be captured to measure the contribution of individual projects.

Third, not all missions may be amenable to this kind of analysis. In the GPS example, simply developing the metric took some time, and even then, it may not capture the full extent of what the GPS program must deliver. Many other missions, from space to cyber to training (of all kinds), may have more difficulty developing mission outcome metrics that are both modelable and communicate in a language understandable by the rest of the Air Force institution.

Fourth, this approach probably excludes many infrastructure assets and activities that may be more distant from individual missions, like functions of the base as a "municipality." In discussions of this sort, the example of child care centers or gymnasiums is often given, but for many facilities, it simply is not clear whether a disruption or degradation would have any quantifiable impact on mission outcomes, though it may have a noticeable impact on personnel disruptions and morale. Thus, this kind of modeling approach simply would not have a way to capture investment in those assets.

Finally, this approach still may require integration across projects and bases that all contribute to the same mission. For example, if a given level of funding is insufficient to support three projects at three bases, all of which affect fighter pilot training, and separate models capture the local impacts, a separate model will probably be necessary to integrate the local impacts into an enterprise-level impact to the mission area.

Mitigating the Weaknesses

All that said, some of the advantages of this approach may be salvageable. This approach could be taken only for missions for which it is otherwise hard to envision the effect of infrastructure. The GPS model is one example. Or perhaps it could be targeted at certain types of infrastructure. For a sortie generation mission, the impacts of runways and maintenance hangars may be relatable, but other infrastructure systems may not be.

In addition, less sophisticated models could be developed to limit the costliness of development but still explore specific relationships or outcomes. For example, the model we developed in Chapter Four is on the low-cost side, whereas PRAS required many years and millions of dollars to develop and answers a broad array of very complicated questions.

Finally, this kind of modeling could be limited to individual projects that could help communicate their impact. In Chapter Four, we found that the data probably do not exist today to fully tie infrastructure asset funding to mission outcomes. But subject-matter expert input could be used to tie the effects of delaying individual projects.

The use of mission outcome models could be compelling, but more thought and care should go into selecting cases where the conditions are right, and the benefits justify the effort required.

Composite Risk Metrics

Strengths

One of the benefits of using composite risk metrics is that the Air Force can already leverage the existing AF/A9 risk framework. If that risk framework has enough traction in the institution, it could provide language that would already be comprehensible by the AFCS.

Second, the Air Force's data systems (e.g., SMSs such as BUILDER) already have some, if not all, of the data-handling capability needed. They are not all completely populated, but systems like BUILDER are designed to serve functions like this. They already have some of the metrics and sub-elements, like condition and functionality, but also the ability to use thresholds to trigger actions. To the degree that these data can be exported and synthesized, they can be leveraged to tie to a risk framework. This means that once users have the source data, much of the process could be automated.

This approach can also be used to express long-term impacts of underfunding. And the SMSs the Air Force uses already have some ability to create multiyear displays.

Weaknesses

While the AF/A9 risk framework provides a ready-made language to communicate to the AFCS, this is the most abstract of the three approaches we explored. One cannot see and touch stoplight metrics or risk indexes. It is often hard to moderate differences in risk in frameworks like this, and two options that appear to fall into the same category may have important differences.

Second, while the Air Force's data systems provide the structure and machinery to process and output much of the needed data, the data must first be gathered. True, DoD has directed that the services adopt and populate the SMSs that are available (DoD, 2013b). But the guidance specified only that the facility CI must be populated, and other elements of infrastructure performance are important to capture to best tie to mission risk. Further, the systems are not yet fully populated, and incomplete data (i.e., not all bases, not all facilities on each base) could prevent valid analysis from being produced. Beyond fulfilling the letter of the law for DoD guidance, the Air Force faces a question of cost-benefit trade-offs as to how much additional time and effort to put into populating these systems.

As one example of implementation costs, the Army estimated the financial cost to implement its detailed development of MDI (only one element of this approach, besides condition and function) at $40,000–$75,000 per Army installation, which could be several million dollars or more across the Air Force, if applied force-wide.

Finally, using metrics like condition and performance as proxies for mission performance could produce misleading results. Commanders are under pressure to produce success regardless of the challenges. Infrastructure failures (or near-failures) are often overcome by creative leadership and action that avoid the natural consequences of funding decisions.[1] One senior leader jokingly said that commanders often say "If the metric is performance, we're 'green'; if the metric is resources, we're 'red.'" This captures a common concern in both infrastructure and logistics communities.

Yet capability and performance must be measured, and resources must be allocated based on criticality of need. There are a number of ways to overcome this difficulty, such as separating metrics used to judge commanders' performance from those used to allocate resources; judging infrastructure condition and functionality by engineers and other technical personnel using rigorous, transparent methods; or having impartial judges, such as inspectors, assess the veracity of the data. Whatever the method, it must be addressed, and not simply ignored. Besides accurately assessing these metrics to properly guide day-to-day decisions, the credibility of the system rests on high-fidelity data.

Mitigating the Weaknesses

One of the key questions for this approach is how much data (and of what fidelity) is really needed to make it work. There is almost no end to how much data could be gathered. But one of

---

[1] We thank personnel from AFCEC-CP for raising this point (on November 17, 2015).

the keys to make this or any risk communication strategy work is developing credibility in the methods used to develop the analysis. The Air Force as a whole, and particularly the AFCS, must believe that the data are rigorous and objective enough to drive real funding decisions. It is beyond the scope of this project to do such a cost-benefit analysis, but further research could illuminate this.

6. Conclusions and Recommendations

The Air Force faces an enormous challenge in articulating the mission impact of underfunding infrastructure SRM activities. In Chapter One, we laid out some of the obstacles, including structural incentives, decisionmaking biases, and a need to educate stakeholders and decisionmakers. In this report, we reviewed several approaches to articulating the mission impact of infrastructure funding to inform the Air Force's choices. We now summarize a few of our conclusions.

Conclusions

There are several viable approaches the Air Force can take to articulate mission impact; each has very different strengths, weaknesses, and implementation burden. There was no approach to communicating mission risk of underfunding SRM that arose in the literature or our analysis as a "best practice." MCDA methods, described in Chapter Three, are used almost without exception to *prioritize* investments, but the actual "sales pitch" of presenting a funding request to a board or other decisionmaking body is not often discussed. Various fields of risk analysis have tried-and-true methods, but the task of translating the analysis of risk to the communication of that risk appears to be an organization-specific task, and one that most noncommercial organizations struggle with.

That said, all three approaches we reviewed are widely used in public and corporate decisionmaking and in policy analysis. We believe that all three approaches may have a place in the Air Force as it transitions away from the status quo. Choosing a path ahead will require more thought and collaboration with infrastructure users and AFCS decisionmakers, and implementing that approach will likely require gathering more information. This analysis was originally envisioned as a multiyear undertaking, and we make recommendations below as to how the Air Force can proceed in the most cost-effective way.

That said, **the infrastructure-to-mission mapping exercise appears to have several potential side benefits.** In Chapter Four, we used these mappings (i.e., logic models) to develop computer models of mission outcome metrics. But these maps can reveal and clarify critical linkages. It could be useful to incorporate these products in a base's development of its contingency response plan (CRP) requirements (Air Force Instruction 10-211, 1998). We were told that, in many cases, asset-to-asset dependencies are missed, and the mapping can provide more in-depth response measures.[1]

---

[1] Feedback provided by personnel from AFCEC-CP, November 17, 2015.

This could also be leveraged to inform currently implemented metrics, such as MDI. The Navy and others explicitly include "replaceability" in their MDI metric. This characteristic could be revealed or at least justified by creating these mappings. The Air Force has a process to adjudicate MDI changes for infrastructure assets; these mappings could inform that process.

Solid risk analysis and communication are necessary, but not sufficient, for successful advocacy for infrastructure funding in the POM. We foreshadowed this in Chapter One in laying out the widespread difficulties many organizations in the United States have in securing adequate infrastructure funding. In a range of fields—MCDA, risk analysis, risk communication, and infrastructure management, specifically—several themes repeatedly arose: high-level institutional buy-in, education of nontechnical personnel, collaboration and iteration to establish decisionmaking values and criteria, and the importance of developing a robust institutional decisionmaking environment and process.

In light of our conclusions, we offer several recommendations.

Recommendations

Assess the POM environment more deeply to determine the best way to implement the project scorecard approach. Of the three approaches we assessed, this seems to us the only viable one that can be implemented in the near term, to potentially improve on the status quo approach to presenting the POM request. We offered a number of cautions in this approach, but they could be mitigated to offer some additional value by presenting some of the project information in POM deliberations. We believe that the viability of this approach depends in part on the contents of the project tradespace and in part on how the material is presented. The project tradespace can be assessed by the AFIMSC and CE community to see what challenges they might face in developing and presenting POM inputs. Determining how to actually present those results should be done in collaboration with relevant personnel from the AFCS.

We believe that a fresh assessment of the decision environment will help guide the Air Force's way ahead, especially in light of the recent changes in responsibility (including the formation of the AFIMSC) and project prioritization process. Once that assessment is made, the choices may become clearer.

Continue to fully populate existing SMSs, and embrace and implement new ones as they are launched, with an eye toward informing a composite risk metric approach. The Air Force must do this to some degree anyway, in order to meet DoD's recent guidance. At the time of our research, this rollout was just beginning and had progressed well at a few bases but still had a long way to go. We were told that the Air Force is working hard to populate BUILDER and that PAVER is nearly 100 percent complete for airfield pavements. We were also told that data entry errors are already a problem with implementation, and that SMSs are subject to the "garbage in—garbage out" adage of data models.[2]

But how far the Air Force goes in populating these SMSs, beyond the letter of the law, depends on the anticipated payoff. Fortunately, both the Army and Navy appear to be further along in pursuing infrastructure performance and mission metrics, collecting data, and populating their SMSs. The Air Force can consult with the other services (as they are two of the closest analog organizations in size and scope) to see how their own investments have paid off, and hopefully get the best payoff themselves.

Populating SMSs with relevant data has obvious payoffs for near-term infrastructure management (e.g., prioritizing where maintenance should be done at a tactical level). This is one of the original purposes of these information systems. But these systems, as they are more fully populated, can be mined for valuable information to feed higher-level analysis for POM input like the approaches we reviewed in this report. Populating these systems should be a near-term priority, but some of the more powerful analyses will take time as data are gathered over the long term. As these systems become more populated, and the fidelity of the data validated, the potential contribution of the composite risk metrics approach (which depends on these data) can be better assessed.

Make targeted assessments to use models to quantify mission outcome metrics. When done right, mission outcome metrics (and their supporting models) can provide especially compelling results, but they are narrow in scope and can also require significant effort. As a result, their application should be carefully calibrated to the desired outcomes. As the Air Force assesses the challenges associated with the project scorecard approach, there may be cases where the mission outcome models could offer assistance in quantifying the mission contributions of some infrastructure projects.

Unfortunately, the personnel with the most expertise in selecting mission metrics and quantifying them—operators and infrastructure users—do not have time to develop those models. We judge that developing models, as exemplified in Chapter Four, will require the collaboration of mission owners, engineers, and analysts.

Finally, **undertake high-level institutional action to educate stakeholders about the effects of infrastructure underfunding**. The CE community greatly needs mission owners to help articulate the value of infrastructure in supporting Air Force missions and the dangers of infrastructure degradation, and mission owners need the CE community. Further, this probably requires effort outside the normal POM process itself. During the POM process, attention becomes focused on very narrow, near-term objectives, not least of which is simply negotiating an executable POM. This environment is not ideal for opening and educating minds about this admittedly complex topic. There are a number of possible avenues and forums—both formal and informal—in which this education could take place. The CE community should consider as

---

[2] Feedback provided by personnel from AFCEC-CP, November 17, 2015.

broad an approach to this as possible, as the obstacles are bigger than simply understanding the facts.

Future Research

As this was originally envisioned as a multiyear undertaking, there are many possible directions for future research to pursue. One of the most important (and potentially compelling) effects of infrastructure underfunding and degradation that we excluded from our scope is long-term cost growth. There exist several models and methods to estimate these costs. For example, BUILDER has IMPACT, a tool that can do forward-looking predictive analysis of future costs. Besides maintenance backlogs and costs, BUILDER can also estimate reduced service life from building component degradation, which can be used to predict early recapitalization and the concomitant costs.

Another area of research is expressing the effects of underfunding on metrics other than mission impact, strictly speaking. A number of personnel in the CE community raised the concern that while infrastructure is deteriorating, engineers are diligently working to avoid or mitigate mission impacts, using a variety of workarounds. They argue that these (completely justifiable) workarounds could both mask the results of infrastructure underfunding (thus not reflecting in mission outcomes) and lead to higher long-term costs to the Air Force.

Some of these effects could be captured in data analysis—for example, the mix of types of maintenance activities, hands-on maintenance workload trends, and user metrics like performance surveys that capture objective and subjective factors that may not be captured with mathematical models or infrastructure performance indexes as currently implemented. These backward-looking (i.e., historical) data analyses could add fidelity to forward-looking predictions about the future effects of infrastructure funding levels.

Concluding Thoughts

All of the steps we describe will require the Air Force, and the CE community specifically, to invest more time and effort. The challenge they confront is widespread, but no magic bullet exists. Other Air Force communities have also invested significant time and resources over many years in information systems and data to help inform requirements determination and POM advocacy. The operational community uses PRAS; the logistics community uses the Logistics Composite Model, MxCap2, the Aircraft Sustainability Model, and more. The Combat Operations in Denied Environments (CODE) projects at RAND have required significant time and resources to address the challenge of air bases under attack, and much of that time was invested in drawing from a broad community of stakeholders and practitioners to develop consensus on very difficult issues. Given the criticality of infrastructure in the Air Force and the size of the annual investment (though presumably underfunded), it stands to reason that the Air

Force must invest significant time and manpower in developing effective means to analyze and communicate the value of infrastructure funding to senior leaders and decisionmakers.

Appendix. Select Findings from Literature Review of Commercial Approaches to Tracking and Forecasting Condition and Cost

This appendix highlights select findings from commercial practice relevant to Air Force tracking and forecasting condition and cost. The methods and tools the Air Force employs for the collection, storage, and processing of information are critically important for determining the relative success of its infrastructure management initiatives. In industry and academia, particular progress has been made in tracking and forecasting conditions and costs as assets age. Costs here include the costs of maintaining and repairing infrastructure ("agency costs") and reductions in serviceability due to deterioration ("user costs"). One relevant theme in this area over the past several decades has been the development and adoption of software suites for infrastructure maintenance management.

Condition Indexes

We begin with a discussion of metrics used to describe the condition of deteriorating facilities. It is intuitively appealing to talk about a building or runway being in good or poor condition, but it is not obvious how to measure condition in a defensible, repeatable manner that relates to management objectives. Researchers and developers have proposed a large number of competing ways to measure the condition of an asset for many classes of infrastructure. For example, the Pavement Condition Index, Overall Pavement Condition Index, Present Serviceability Index, Pavement Quality Index, Pavement Overall Index, Riding Comfort Index, Surface Distress Index, Structural Adequacy Index, International Roughness Index, and Distress Manifestation Index have all been proposed as indexes for measuring the condition of pavement. Many of these are in use around the world today. However, different indexes have rated the relative health of test sections of pavement differently, so the choice of condition index to use is not an insignificant matter (Gharaibeh, Zou, and Saliminejad, 2009). The pavement condition indexes mentioned are composite CIs, aggregating data on specific distresses like the extent and depth of cracks on the surface of a section of pavement.

As noted above, a variety of performance indicators can be considered for different assets. Generally, these will fall into four categories (Uddin, Hudson, and Haas, 2013, p. 209):

- service and user rating
- safety and sufficiency
- physical condition
- structural integrity/load-carrying capacity.

The first of these will largely be taken from the user perspective. There are likely to be many performance indicators even within this category. For example, pavement engineers use different "roughness" or "riding comfort" measures that summarize the overall ride quality offered by pavement. Performance indicators related to structural integrity, or otherwise related to physical condition, are often the responsibility of engineers to establish and measure. A comprehensive set of performance indicators will include input from both communities.[1]

High-level decisionmakers will be interested in the overall condition of the assets they manage and will thus likely need to aggregate metrics to arrive at composite CIs. The danger here is that aggregation leads to information loss. High-level decisions made based on composite CIs may not reflect engineering best practice at the facility level, leading to unnecessary maintenance and repair activities, decreased serviceability, or both simultaneously (Kuhn, 2011).

It can be expensive and time-consuming to calculate the conditions of assets. There are proxy metrics that can be used to easily and quickly gain (limited) insight into conditions. An example would be to look at the age of assets, with the implicit assumption being that older assets are in worse condition. Other proxy metrics include the time that has passed since an asset was last inspected or repaired, or the remaining functional life of an asset. These metrics are clearly not as informative as the results of detailed condition surveys.

The challenges facing the Air Force require identifying multiple indexes with which to track the conditions of managed assets, recognizing the utility and limitations of each. Proxy metrics such as asset age will be required for assets that are difficult or expensive to inspect. Metrics focused on physical condition will be necessary, particularly for assets where there is some risk of structural failure and resulting mission impacts. Serviceability metrics will be important, particularly for assets that are used regularly and whose mission support performance varies over time.

Cost Estimation

The software systems mentioned above frequently select suggested maintenance schedules for assets by minimizing "costs," where costs comprise the maintenance and repair costs associated with maintaining the conditions of assets above predefined limits (agency costs), user costs (which are inversely proportional to the level of service that the assets offer to users), or some combination of the two. The split between agency and user costs arises from the fact that infrastructure maintenance is often done by public agencies managing facilities used by the public at large and charged with maximizing social welfare. In addition, the assets being managed are often facilities like roads or bridges, where many facilities are usable but differ dramatically in terms of how structurally sound they are or how easy, efficient, safe, and/or costly the facilities are to use. Using a cost-minimizing framework requires being able to forecast

---

[1] See Uddin, Hudson, and Haas, 2013, Table 8-2, pp. 210–212, for a large list of examples of performance indicators across multiple asset classes.

costs as a function of maintenance management policy. The need to forecast future costs, as well as the desire to track and analyze past performance, requires tracking costs. Thus, it is common in industrial practice and academic research to track and extrapolate user and agency costs.

Markow (1990) is one of the earliest of many studies to collect and compare agency and user costs, suggesting using life-cycle cost analysis to guide maintenance decisionmaking and highlighting how effective many maintenance activities look when using this type of analysis. Khurshid et al. (2011) and other studies investigate the cost-effectiveness of specific maintenance activities. Khurshid et al. investigated a handful of common rehabilitation activities performed on rigid pavement. The availability of results in prior published research that describe costs as a function of asset condition vary by asset type, with, again, pavement being relatively well understood. Ben-Akiva and Gopinath (1995) introduced a general methodology for estimating the user costs associated with railroad tracks, bridges, highways, and other infrastructure assets, then applied the methodology in the context of Brazilian highways. In an article titled "Expected Maintenance Costs of Deteriorating Civil Infrastructures," Frangopol and Kong (2001) introduced another general framework, focused on estimating the number of maintenance activities and the agency costs that a deteriorating facility will require.

There is a special case involving infrastructure assets that are capacitated—only capable of serving a limited number of users. Often, the number of users that the assets can serve depends on the condition of the assets. In other cases, performing maintenance or repair activities requires suspending use of assets for some time. In either situation, capturing the costs and benefits of maintenance and repair activities requires estimating the value of being able to use a facility or the cost of not being able to use a facility. For example, Zou and Madanat (2011) look at the management of runways at busy airports. Performing a maintenance activity at a runway will prohibit the use of that runway for some time. Separation standards between runway operations mean that there are only so many aircraft takeoffs and landings that a runway or set of runways can accommodate in a set amount of time. During maintenance and repair activities on one runway, aircraft that would have used that runway must instead use other runways. In cases where maintenance must be done during times when demand for runway use exceeds supply, the result is delays that create sizable economic costs (though many runways could support significantly higher utilization). (Clearly, the preferred option would be to perform maintenance activities when there is minimal demand and/or alternate facilities can handle whatever traffic/demand would normally be serviced by facilities being maintained.)

Lists of Figures and Tables

Figures

Tables

Abbreviations

AETC	Air Education and Training Command
AFB	Air Force Base
AFCAMP	Air Force Comprehensive Asset Management Plan
AFCEC	Air Force Civil Engineering Center
AFCS	Air Force Corporate Structure
AFIMSC	Air Force Installation and Mission Support Center
AFPM	Air Force Policy Memorandum
BCAMP	Base Comprehensive Asset Management Plan
BCE	base civil engineer
C2	command and control
CAMP	Comprehensive Asset Management Plan
CARA	Critical Asset Risk Assessment
CARM	Critical Asset Risk Management
CATCODE	Category Code
CE	civil engineering
CI	Condition Index
CIP	Critical Infrastructure Program
DCIP	Defense Critical Infrastructure Program
DoD	U.S. Department of Defense
ERD	estimated range deviation
ERDC-CERL	Engineer Research and Development Center, Construction Engineering Research Laboratory
FCI	Facility Condition Index
FFI	Facility Functionality Index
FI	Functionality Index
FSM	Facility Sustainment Model
FY	fiscal year
FYDP	Future Years Defense Plan
GAO	U.S. Government Accountability Office; U.S. General Accounting Office
GBP	pounds, Great Britain
GPS	Global Positioning System
ID	identity
IFOM	installation figure of merit
IMPACT	Integrated Multiyear Prioritization and Analysis Tool

IPL	Integrated Priority List
ISO	International Organization for Standardization
JCS	Joint Chiefs of Staff
LADWP	Los Angeles Department of Water and Power
M&R	maintenance and repair
MAJCOM	major command
MCA	Multi-Criteria Analysis
MCDA	Multi-Criteria Decision Analysis
MCDM	Multi-Criteria Decision Making
MDI	Mission Dependency Index
MODA	Multi-Objective Decision Analysis
MR&R	maintenance, renovation, and reconstruction
MTBCF	mean time between critical failure
MxCap2	maintenance capability and capacity
NASA	National Aeronautics and Space Administration
NRC	National Research Council
PAF	RAND Project AIR FORCE
POM	Program Objective Memorandum
PRAS	Predictive Readiness Assessment System
R&D	research and development
R&M	repair and maintenance
RPCS	Real Property Categorization System
SIR	savings-to-investment ratio
SMS	Sustainment Management System
SORTS	Status of Resources and Training System
SRM	sustainment, restoration, and modernization
SUPT	Specialized Undergraduate Pilot Training
TAB-VAM	Theater Airbase Vulnerability Assessment Model

References

Abel, Robert, "DoD to Develop Vulnerability Scorecard for Weapons Systems and More," *SC Magazine*, September 21, 2015.

Air Force Doctrine Document 1, *Air Force Basic Doctrine, Organization, and Command*, Washington, D.C.: Department of the Air Force, October 14, 2011.

Air Force Instruction 10-211, *Civil Engineer Contingency Response Plan*, Washington, D.C.: Department of the Air Force, July 1, 1998.

Air Force Instruction 16-501, *Operations Support: Control and Documentation of Air Force Programs*, Washington, D.C.: Department of the Air Force, August 15, 2006.

Air Force Instruction 32-1032, *Civil Engineering: Planning and Programming Appropriated Funded Maintenance, Repair, and Construction Projects*, Washington, D.C.: Department of the Air Force, October 17, 2014.

Air Force Policy Memorandum to Air Force Policy Directive 10-24, *Air Force Critical Infrastructure Program (CIP)*, Washington, D.C.: Department of the Air Force, January 6, 2012.

Alter, C., and S. Murty, "Logic Modeling: A Tool for Teaching Practice Evaluation," *Journal of Social Work Education*, Vol. 33, No. 1, 1997.

American Society of Civil Engineers, *2013 Report Card for America's Infrastructure*, 2013.

Bazerman, M. H., *Judgement in Managerial Decision Making*, 4th ed., New York: John Wiley, New York, 1998.

Ben-Akiva, Moshe, and Dinesh Gopinath. "Modeling Infrastructure Performance and User Costs," *Journal of Infrastructure Systems*, Vol. 1, No. 1, March 1995, pp. 33–43.

CBRE Whitestone, "Estimates of Unscheduled Facility Maintenance," June 12, 2015. As of October 3, 2016:
http://www.cbre.us/o/facility-cost-forecasting/teams/facility-cost-forecasting/research/Pages/Estimates-of-Unscheduled-Facility-Maintenance.aspx

Chesler, Leonard G., and B. F. Goeller, *The STAR Methodology for Short-Haul Transportation: Transportation System Impact Assessment*, Santa Monica, Calif.: RAND Corporation, R-1359-DOT, 1973. As of October 3, 2016:
http://www.rand.org/pubs/reports/R1359.html

Davis, Paul K., *Analytical Architecture for Capabilities-Based Planning, Mission-System Analysis, and Transformation*, Santa Monica, Calif.: RAND Corporation, MR-1513-OSD, 2002. As of October 3, 2016:
http://www.rand.org/pubs/monograph_reports/MR1513.html

Davis, Paul K., Russell D. Shaver, and Justin Beck, *Portfolio-Analysis Methods for Assessing Capability Options*, Santa Monica, Calif.: RAND Corporation, MG-662-OSD, 2008. As of October 3, 2016:
http://www.rand.org/pubs/monographs/MG662.html

Davis, Paul K., Stuart E. Johnson, Duncan Long, and David C. Gompert, *Developing Resource-Informed Strategic Assessments and Recommendations*, Santa Monica, Calif.: RAND Corporation, MG-703-JS, 2008. As of October 3, 2016:
http://www.rand.org/pubs/monographs/MG703.html

De Sitter, W.R., "Costs for Service Life Optimization: The "Law of Fives," in Steen Rostam, ed., *Durability of Concrete Structures Workshop Report*, ed., Copenhagen, Denmark, 1984, pp. 131–134.

Defense Business Practice Implementation Board, *Implementation of Balanced Scorecard Metrics*, Report FY02-2, 2002.

Department of Defense Directive 3020.40, *Defense Critical Infrastructure Program (DCIP)*, Washington, D.C.: U.S. Department of Defense, 2010.

Department of Defense Instruction 3020.45, *Defense Critical Infrastructure Program (DCIP) Management*, Washington, D.C.: U.S. Department of Defense, 2008.

Department of Defense Instruction 4165.03, *DoD Real Property Categorization*, Washington, D.C.: U.S. Department of Defense, February 4, 2015.

DoD—*See* U.S. Department of Defense.

Dunn, Steven C., and Melvin H. Sawyer, "A Proposed Approach for Prioritizing Maintenance at NASA Centers," paper presented at the meeting of the American Institute of Aeronautics and Astronautics, Reston, Va., January 7–10, 2013. As of October 3, 2016:
https://ntrs.nasa.gov/search.jsp?R=20130003314

Durango, Pablo L., and Samer M. Madanat, "Optimal Maintenance and Repair Policies in Infrastructure Management Under Uncertain Facility Deterioration Rates: An Adaptive Control Approach," *Transportation Research Part A: Policy and Practice*, Vol. 36, No. 9, November 2002, pp. 763–778.

Frangopol, Dan M., Jung S. Kong, and Emhaidy S. Gharaibeh, "Reliability-Based Life-Cycle Management of Highway Bridges," *Journal of Computing in Civil Engineering*, Vol. 15, No. 1, 2001, pp. 27–34.

Gallagher, Mark A., Douglas A. Boerman, Cameron A. MacKenzie, and David M. Blum, "Improving Risk Assessment Communication," *Military Operations Research Society Journal*, Vol. 2, No. 1, 2016.

GAO—*See* U.S. Government Accountability Office.

Gharaibeh, Nasir G., Yajie Zou, and Siamak Saliminejad, "Assessing the Agreement Among Pavement Condition Indexes," *Journal of Transportation Engineering*, Vol. 136, No. 8, 2009, pp. 765–772.

Goeller, B. F., Allan Abrahamse, James H. Bigelow, Joseph G. Bolten, David M. De Ferranti, James C. DeHaven, T. F. Kirkwood, and Robert Petruschell, *Protecting an Estuary from Floods—A Policy Analysis of the Oosterschelde*, Vol. 1, *Summary Report*, Santa Monica, Calif.: RAND Corporation, R-2121/1-NETH, 1977. As of October 3, 2016: http://www.rand.org/pubs/reports/R2121z1.html

Gordon, Larry, "UCLA Claims $13 Million in Flood Damage from Water Line Break," *Los Angeles Times*, July 9, 2015.

Grussing, Michael N., Steve Gunderson, Mary Canfield, Ed Falconer, Albert Antelman, and Samuel L. Hunter, *Development of the Army Facility Mission Dependency Index for Infrastructure Asset Management*, Vicksburg, Miss.: U.S. Army Corps of Engineers, Engineering Research and Development Center, September 2010a.

Grussing, Michael N., Lance R. Marrano, and Matthew C. Walters, *Development of Army Facility Functionality Assessment Criteria and Procedures*, No. ERDC/CERL-TR-10-17, Champaign, Ill.: Engineer Research and Development Center, Construction Engineering Research Lab, 2010b.

Halfawy, M. R., Linda A. Newton, and Dana J. Vanier, "Review of Commercial Municipal Infrastructure Asset Management Systems," *Electronic Journal of Information Technology in Construction*, Vol. 11, 2006, pp. 211–224.

Harrison, Todd, "Rethinking Readiness," *Strategic Studies Quarterly*, Vol. 8, No. 3, Fall 2014, pp. 38–68.

Hastings, Nicholas A. J., *Physical Asset Management*, New York: Springer, 2010.

Hillestad, Richard, and Paul K. Davis, *Resource Allocation for the New Defense Strategy: The DynaRank Decision-Support System*, Santa Monica, Calif.: RAND Corporation, MR-996-OSD, 1998. As of October 3, 2016: http://www.rand.org/pubs/monograph_reports/MR996.html

Institution of Civil Engineers, *Realizing a World Class Infrastructure: ICE's Guiding Principles of Asset Management*, 2013.

International Organization for Standardization, International Standard 55000, *Asset Management—Overview, Principles and Terminology*, January 15, 2014.

Iseler, Tracy K., "Balanced Scorecard Domain School," Office of the Secretary of Defense, Logistics Systems Management, May 1, 2003. As of October 3, 2016: https://acc.dau.mil/adl/ en-US/46162/file/13702/Balanced%20Scorecard%20Domain%20School.pdf

Joint Publication 1-02, *Dictionary of Military and Associated Terms*, Washington, D.C.: U.S. Joint Chiefs of Staff, 2005.

Kaplan, R. S., and D. P. Norton, "The Balanced Scorecard: Measures That Drive Performance," *Harvard Business Review*, January–February 1992.

Kaplan, R. S., and D. P. Norton, *The Balanced Scorecard: Translating Strategy into Action*, Cambridge, Mass.: Harvard Business Review Press, September 1, 1996.

Keeney, Ralph L., and Howard Raiffa, Cambridge, UK: Cambridge University Press, *Decisions with Multiple Objectives: Preferences and Value Tradeoffs*, 1976.

Khurshid, Muhammad Bilal, Muhammad Irfan, and Samuel Labi. "An Analysis of the Cost-Effectiveness of Rigid Pavement Rehabilitation Treatments," *Structure and Infrastructure Engineering*, Vol. 7, No. 9, 2011, pp. 715–727.

Knowles, Scott Gabriel, "Learning from Disaster? The History of Technology and the Future of Disaster Research," *Technology and Culture*, Vol. 55, No. 4, October 2014, pp. 773–784.

Kuhn, Kenneth D., "Pavement Network Maintenance Optimization Considering Multidimensional Condition Data," *Journal of Infrastructure Systems*, Vol. 18, No. 4, 2011, pp. 270–277.

Lufkin, P., A. Desai, and J. Janke, "Estimating the Restoration and Modernization Costs of Infrastructure and Facilities," *Public Works Management and Policy*, July 2005.

Markow, Michael J., "Life-Cycle Cost Evaluations of the Effects of Pavement Maintenance," *Transportation Research Record*, Vol. 1276, 1990, pp. 37–47.

McNerney, Michael T., and Robert Harrison, "Full-Cost Approach to Airport Pavement Management," in *Handbook of Airline Economics*, New York: Aviation Week Group, 1995, pp. 121–129.

Mizusawa, Daisuke, "Road Management Commercial Off-the-Shelf Systems Catalog," University of Delaware, February 2009.

Narasimhan, S., and Wallbank, E. J., *Bridge Inspections: Current Thinking and Regimes*, 8th International Conference on Extending the Life of Bridges, Civil and Building Structures, Engineering Technics Press, July 1999.

National Council on Public Works Improvement, *Fragile Foundations: A Report on America's Public Works*, Washington, D.C., 1988.

National Research Council, *Committing to the Cost of Ownership: Maintenance and Repair of Public Buildings*, Washington, D.C.: National Academy Press, 1990.

———, *Stewardship of Federal Facilities: A Proactive Strategy for Managing the Nation's Public Assets*, Washington, D.C.: National Academy Press, 1998.

———, *Investments in Federal Facilities: Asset Management Strategies for the 21st Century*, Washington, D.C.: National Academies Press, 2004.

———, *Predicting Outcomes from Investments in Maintenance and Repair for Federal Facilities,* Washington D.C.: National Academies Press, 2012.

NRC—*See* National Research Council.

Raiffa, Howard, *Preferences for Multi-Attributed Alternatives*, Santa Monica, Calif.: RAND Corporation, RM-5868-DOT/RC, 1969. As of October 3, 2016:
http://www.rand.org/pubs/research_memoranda/RM5868.html

Regan, Edward V., "Holding Public Officials Accountable for Infrastructure Maintenance," *Proceedings of the Academy of Political Science*, Vol. 37, No. 3, 1989, pp. 180–186.

Reid, Robert L., "Special Report: The Infrastructure Crisis," *Civil Engineering*, Vol. 78, No. 1, January 2008.

Secretary for the Air Force, Deputy Assistant Secretary for Budget, *United States Air Force: Fiscal Year 2015 Budget Overview*, U.S. Air Force, March, 2014. As of April 6, 2016:
http://www.saffm.hq.af.mil/shared/media/document/AFD-140304-039.pdf

Sitzabee, William E., and Marie T. Harnly, "A Strategic Assessment of Infrastructure Asset-Management Modeling," *Air and Space Power Journal*, Vol. 27, No. 6, November–December 2013.

Snyder, Don, Patrick Mills, Katherine Comanor, and Charles Robert Roll, Jr., *Sustaining Air Force Space Systems: A Model for the Global Positioning System*, Santa Monica, Calif.: RAND Corporation, MG-525-AF, 2007. As of October 3, 2016:
http://www.rand.org/pubs/monographs/MG525.html

Streicher, Burton L., *Navy Shore Infrastructure Investment Support: Modeling of Investment to Output Performance and Readiness*, Alexandria, Va.: CNA, November 2008.

Surowiecki, James, "System Overload," *New Yorker*, April 18, 2016.

Thomas, Brent, Mahyar A. Amouzegar, Rachel Costello, Robert A. Guffey, Andrew Karode, Christopher Lynch, Kristin F. Lynch, Ken Munson, Chad J. R. Ohlandt, Daniel M. Romano, Ricardo Sanchez, Robert S. Tripp, and Joseph V. Vesely, *Project AIR FORCE Modeling Capabilities for Support of Combat Operations in Denied Environments*, Santa Monica, Calif.: RAND Corporation, RR-427-AF, 2015. As of October 3, 2016: http://www.rand.org/pubs/research_reports/RR427.html

Uddin, Waheed, W. Ronald Hudson, and Ralph Haas, *Public Infrastructure Asset Management, 2nd Ed.*, New York: McGraw-Hill Education, 2013.

United Kingdom Department for Communities and Local Government, *Multi-Criteria Analysis: A Manual*, January 2009.

U.S. Air Force, *FY17–18 AFCAMP Playbook Consolidated PDF*, Washington, D.C., August 14, 2015.

U.S. Department of Defense, *DoD Mission Assurance Strategy*, Washington, D.C., April 2012.

———, *DoD Base Structure Report, Fiscal Year 2013*, Washington, D.C., 2013a.

———, *Memorandum: Standardizing Facility Condition Assessments*, Washington, D.C., September 10, 2013b.

U.S. Government Accountability Office, *Transportation Infrastructure: Limited Improvement in Bridge Conditions over the Past Decade, but Financial Challenges Remain*, Washington, D.C., GAO-13-713T, 2013.

———, *High-Risk Series: An Update*, Washington, D.C., GAO-15-290, February 2015.

U.S. Navy, Naval Facilities Engineering Service Center, *Mission Dependency Index (MDI): An Operational Risk Metric for Assessing the Criticality of Naval Shore Facilities*, Port Hueneme, Calif.: Naval Facilities Engineering Command, n.d.

Vanier, D. J., "Why Industry Needs Asset Management Tools," *Journal of Computing in Civil Engineering*, Vol. 15, No. 1, 2001, pp. 35–43.

Whitestone Research and Jacobs Facilities Engineering, *Implementation of the Department of Defense Sustainment Model: Final Report*, Washington, D.C., January 2001.

Zou, Bo, and Samer Madanat, "Incorporating Delay Effects into Airport Runway Pavement Management Systems," *Journal of Infrastructure Systems*, Vol. 18, No. 3, 2011, pp. 183–193.